北京林业大学习近平生态文明思想研究专项（项目编号：2021STWM01）

理地营境
生态文明建设背景下风景园林实践

LAND DESIGN AND ARTISTIC PERFECTION
The Practice of Landscape Architecture in the
Context of Ecological Civilization

李雄　郑曦　李运远　著

中国建筑工业出版社

图书在版编目（CIP）数据

理地营境：生态文明建设背景下风景园林实践＝LAND DESIGN AND ARTISTIC PERFECTION The Practice of Landscape Architecture in the Context of Ecological Civilization / 李雄，郑曦，李运远著. -- 北京：中国建筑工业出版社，2021.11
ISBN 978-7-112-26792-7

Ⅰ. ①理… Ⅱ. ①李… ②郑… ③李… Ⅲ. ①生态环境建设②园林设计 Ⅳ. ①X321.2②TU986.2

中国版本图书馆CIP数据核字（2021）第211107号

责任编辑：杜　洁　李玲洁
责任校对：姜小莲

理地营境

生态文明建设背景下风景园林实践
LAND DESIGN AND ARTISTIC PERFECTION
The Practice of Landscape Architecture in the Context of Ecological Civilization
李雄　郑曦　李运远　著
*
中国建筑工业出版社出版、发行（北京海淀三里河路9号）
各地新华书店、建筑书店经销
北京锋尚制版有限公司制版
天津图文方嘉印刷有限公司印刷
*
开本：965毫米×1270毫米　1/16　印张：15¾　字数：323千字
2022年9月第一版　　2022年9月第一次印刷
定价：**169.00**元
ISBN 978-7-112-26792-7
　　　（38082）

版权所有　翻印必究
如有印装质量问题，可寄本社图书出版中心退换
（邮政编码100037）

前　言

当前中国特色社会主义进入新时代，我国城市生态和人居环境面临着新的形势和全新挑战。党的十九届五中全会将"生态文明建设实现新进步"作为"十四五"时期经济社会发展的主要目标之一。党的十八大以来，以习近平同志为核心的党中央将"生态文明"建设纳入中国特色社会主义"五位一体"总体布局和"四个全面"战略布局，引起我国现代化建设中各行业对自身角色和任务的重新思考。风景园林学作为人居环境科学的支撑学科，承担着协调人类与自然的关系、建设美丽中国的任务，在中国新时代生态文明建设的工作中肩负着重大的责任和历史使命。

一、习近平新时代生态文明建设的新思想

习近平新时代生态文明思想是中国特色社会主义思想的重要组成部分，也是指导我国绿色发展的科学指南。2018年2月，习近平总书记在视察成都天府新区时首次提出"公园城市"理念，指示"天府新区一定要规划好建设好，特别是要突出公园城市特点，把生态价值考虑进去"。2018年5月，习近平总书记在全国生态环境保护大会上发表重要讲话，对推进新时代生态文明建设提出必须遵循六项重要原则：把"坚持人与自然和谐共生"放在首位；用"绿水青山就是金山银山"的思想处理发展与保护的关系；强调"良好生态环境是最普惠的民生福祉"和"山水林田湖草是生命共同体"；用"最严格制度、最严密法治"为生态文明建设保驾护航和共谋全球生态文明建设；深度参与全球环境治理，推动构建人类命运共同体。习近平总书记对生态文明建设和生态环境保护提出一系列新理念、新思想、新战略，为新时代生态环境保护工作提供了重要指引和根本遵循，同时也对我国公园城市建设和城市人居生态环境发展指出了新的发展方向和建设目标。

二、生态文明背景下风景园林的发展内涵

1. 积极落实"以人民为中心"的思想

建设生态文明，关系人民福祉，关乎民族未来。良好的生态环境是最公平的公共产品，是最普惠的民生福祉。风景园林的发展和实践应该坚持以"以人民为中心"的发展思维，通过打造人人可享受的高品质城乡人居环境以满足人民日益增长的美好生活需求，打造开放性、可达性、亲民性的公园体系，提升城市综合竞争力。公园城市建设的核心在于"公"，面向公众，公平共享。因此，公园城市建设也应从为城市居民服务的角度，实现绿色空间生态、景观、游憩、文化、科教、防灾等多种功能的协调发展，切实提高人民群众的幸福感和获得感。

2. 构建人与自然和谐共生的理想蓝图

构建人与自然和谐共生的生态自然观是生态文明建设的思想基础，同时也是风景园林实现"协调人与自然关系"根本任务的理论基础。突破"公园"的单一概念，建立"山水林田湖草生命共同体"综合体系。风景园林要在城市建设用地内的公园体系营造基础上，将人居生态环境有机整合，拓展对于自然环境、区域环境、人工环境的更大范围、更多层次的系统性思考。积极融入"山水林田湖草是生命共同体"的习近平生态文明思想，建立"城市—乡村—自然保护地"的公园城市综合体系，形成"人与自然和谐共生的大系统"，这是适应生态文明理念下人居生态环境营造的必然，也将有效推动公园城市在我国生态文明建设中发挥更为积极的作用。

3. 形成交叉多元融合的学科理论体系

　　面对新时代生态文明建设的形势和挑战，在公园城市建设的实践中，风景园林应该融合交叉多学科理论与技术，强化人居环境学科和其他相关学科的融合，在风景园林学、城乡规划学和建筑学的基础上，不断结合林学、生态学、地理学等其他学科的理论、方法与技术创新，形成适应新时代发展需求的公园城市建设科学理论体系，奠定公园城市发展基础和技术体系，逐步建立习近平新时代生态文明思想下公园城市建设的科学理论和方法。

4. 继承发展中国传统园林文化

　　新时代生态文明的建设仍植根于中华悠久的传统文化土壤，依靠中国传统的生态智慧，平衡物质文明与精神文明，形成具有中国特色的可持续发展生态理论体系和生态建设模式。在公园城市的建设中强调"天人合一"的中国传统园林理念，将"生境、画境、意境"高度统一，推动人工环境、自然环境和文化底蕴的协调发展。通过构建融入山水自然、彰显中国传统文化特色的城市绿色格局，实现"望山见水记乡愁"，构建诗意栖居的城市理想境界。

三、风景园林助力生态文明建设

1. 做好顶层设计

　　风景园林行业应主动介入城市未来的发展，积极响应人民的需求，解决城市的问题，充分发挥作为绿色基础设施的综合效能。在新形势和新问题的导向下，城市风景园林工作应做好顶层设计，打破单纯服从城市总体规划"用地填空"的被动局面，合理引导空间布局，控制人工建设用地的无序蔓延和肆意扩张。同时跳开城市建设用地的限制，积极整合城市周边自然环境和乡村绿色空间，最终构建一个既能实现生态系统价值和功能，也能为市民提供多种服务功能的公共服务载体和绿色发展支撑平台。

2. 推动公众参与

　　生态文明建设的核心是"以人民为中心"，其建设目标也是满足人民群众对美好生活的需求。风景园林的各项工作应紧密围绕这个核心，一方面，要提高人民群众对环境保护的意识，推动生产生活的绿色发展；另一方面，要关注市民对环境的需求和感受，建设高品质的人居环境。重视城市绿地空间布局的合理性，提高绿地服务的功能和质量，园林规划和建设程序信息公开透明，提高公众对绿地建设的参与度和积极性，注重公众对城市绿化的满意度，以保障城市建设反映民意。

3. 建立弹性机制

　　城市处于不断的发展变化中，尤其在我国当今快速发展的时期，生态文明建设的工作也在环境弹性变化的基础上作出相应的弹性响应，包括生态系统不同阶段的适应、不同对象需求的适应，以及不同尺度范围、不同目标任务的适应等。风景园林需要采用一个动态的、可塑性强的规划方法作为指导，建构一个应对这种日益加剧的动态变化的综合发展战略框架，将可能的动态因素纳入生态文明建设布局中，以适应城市社会经济的发展变化。增强城市绿地的弹性适应力，以此来化解建设与保护、近期与远期之间的矛盾，满足城市良性持续发展的需求。

4. 创造协同平台

　　生态文明建设离不开风景园林、规划、林业、水利、经济、交通等众多专业层面的紧密配合，也需要从政府、企业到普通民众的多方力量的协作。城市园林管理部门应该主动搭建协同平台，构建一个多部门协调合作体系，打破各部门的职能权限局限。通过多方力量形成联合协作的态势，让不同的部门和专业从不同角度为生态文明建设提供科技支撑和技术支持，以解决生态文明建设中遇到的问题，共同推动生态文明，建设美丽中国。

四、展望

　　风景园林作为我国生态和人居环境建设中的核心内容，也必然是引领我国生态文明建设的重要支撑。在"十四五"的开局之年，也是奔向 2035 年远景目标的新起点上，风景园林应以建设美丽中国为己任，主动作为，积极响应，丰富理论基础，培养创新思维，提高综合能力，为我国生态文明建设不断提供理论、方法与技术支撑，做生态文明建设的先行者，开启新时代风景园林的新篇章，迎接社会主义生态文明新时代的到来。

Foreword

As China's socialism has entered a new era, its urban ecology and human settlements are facing new situations and new challenges. The Fifth Plenary Session of the 19th Central Committee of the Communist Party of China regards "new progress in the construction of ecological civilization" as one of the main goals of economic and social development during the 14th Five Year Plan period. Since the 18th Party's Congress, the Central Committee of the CPC, with President Xi Jinping as its core, has brought the "ecological civilization" into the Five-sphere Integrated Overall Plan and the Four-pronged Comprehensive Strategy, which has aroused the new thinking of various industries in China's modernization drive on their roles and tasks. As a supporting discipline of sciences of human settlements, landscape architecture undertakes the task of coordinating the relationship between man and nature and building a beautiful China. It also shoulders great responsibility and historical mission in the development of ecological civilization in the new era of China.

1 President Xi Jinping's new thought on the ecological civilization in the new era

President Xi Jinping's thought on ecological civilization in the new era is an important part of socialism with Chinese characteristics and a scientific guide for China's green development. In February 2018, President Xi Jinping first put forward the concept of "Park City" during his visit to Tianfu New District of Chengdu, and indicated that Tianfu New District must be systematically planned and developed, highlighting the characteristics of "Park City and ecological value". In May 2018, President Xi Jinping delivered an important speech at the National Conference on ecological environment protection. He proposed six important principles for advancing the development of ecological civilization in the new era: putting the harmonious coexistence between man and nature in the first place; dealing with the relationship between development and protection with the vision that lucid waters and lush mountains are invaluable assets; emphasizing that "good ecological environment is the most inclusive people's well-being" and "mountains, rivers, forests, fields, lakes and grass are a community of life"; sustaining national ecological civilization with the strictest rule of law and pushing forward global ecological civilization with joint efforts; deep involvement in global environmental governance to facilitate the development of a community with shared future for mankind. President Xi Jinping put forward a series of new ideas and new strategies for ecological civilization and ecological conservation, which provided important guidelines and fundamental rules for the protection of ecological environment in the new era, and pointed out the new development orientation and development goals for Park City and City Habitat of China.

2 The development connotation of landscape architecture in the context of ecological civilization

1. Actively implementing the people-centered approach

The development of ecological civilization is related to the well-being of the people and the future of the nation. A good ecological environment is the fairest public product and the most inclusive well-being of people's livelihood. The development of landscape architecture and its application should follow a people-centered approach. It should help to improve the comprehensive competitiveness of the city by creating a high-quality urban and rural living environment that everyone can enjoy to meet the people's growing demand for a better life and creating an open, accessible and people-friendly park system. The core of park city development is open to and shared by the "public". Therefore, the park city development should also serve the urban residents, realize the coordinated development of green space ecology, landscape, recreation, culture, science and education, disaster prevention and other functions, and effectively improve the people's sense of happiness and contentment.

2. Building an ideal blueprint for harmonious coexistence between man and nature

The ideological basis for the development of ecological civilization is the establishment of an eco-nature concept of harmonious coexistence between man and nature, which is also the theoretical basis for landscape architecture to realize its fundamental task of "coordinating the relationship between man and nature". Landscape architecture should break the single concept of "park" and establish a comprehensive system of landscape, forest, field, lake and grass. Based on the park space environment in urban construction land, landscape architecture should intensively integrate ecological human settlements and expand the systematic thinking of natural, regional and artificial environments in a larger scope and at more levels. Landscape architecture should also actively integrate President Xi Jinping's idea of ecological civilization that mountains, waters, forests, farmlands, lakes and grasslands are part of a community of life, and establish a comprehensive "city-country-nature reserve" park city system, so as to build a large system of harmonious coexistence between man and nature.

This is an inevitable choice for the development of ecological human settlements in the context of ecological civilization, and it will make park city to play a more positive role in the development of ecological civilization in China.

3. Forming a theoretical system of interdisciplinary and pluralistic integration

Facing the situations and challenges of ecological civilization development in the new era, in the practice of park city construction, landscape architecture should integrate interdisciplinary theory and technology, strengthen the integration of human settlements and other related disciplines, and constantly combine the theory, method and technology of forestry, ecology, geography and other disciplines on the basis of landscape architecture, urban and rural planning and architecture. We should create a scientific theoretical system for park city construction to meet the needs of new era, lay the foundation and establish a technological system for park city development, and gradually establish the scientific theory and method of park city construction under the new era of President Xi

Jinping's ecological civilization.

4. Inheritance and development of the Chinese traditional garden culture

The construction of ecological civilization in the new era is still rooted in China's long-standing traditional cultural soil. Relying on China's traditional ecological wisdom, balancing material civilization and spiritual civilization, it forms an ecological theoretical system and ecological construction mode of sustainable development with Chinese characteristics. In the construction of modern park city, the traditional Chinese gardening concept of "harmony between man and nature" is emphasized, which highly unifies "habitat, painting and artistic conception", and promotes the coordinated development of artificial environment, natural environment and cultural heritage. Through the construction of the urban green pattern integrating the natural landscape and highlighting the characteristics of Chinese traditional culture, we can realize the ideal state of "nostalgia aroused by the scenery of mountain and water" and build a poetic city.

3 Landscape architecture leads the construction of ecological civilization

1. Top-level design

Landscape architecture profession should actively participate in the future development of cities, respond to the needs of the people, solve urban problems and give full play to the comprehensive efficiency of green infrastructure. Oriented towards the new situation and new problems, the top-level design of urban landscape architecture should be done well to break the passive situation of simply obeying the urban plan of "land filling in the void", reasonably guide the spatial layout, and control the disorderly spread and wanton expansion of artificial construction land. At the same time, we should break away from the restriction of urban construction land, actively integrate the natural environment around the city and rural green space, and finally build a medium for public service and a supporting platform for green development that can not only realize the value and function of the ecosystem, but also provide diverse service functions for citizens.

2. Public participation

The core of ecological civilization construction is "people-centeredness", and its construction goal is to meet people's needs for a better life. The work of landscape architecture should be closely around this core. On one hand, it is necessary to improve people's awareness of environmental protection and promote the green development of production and life; on the other hand, it is necessary to pay attention to people's needs and feelings for the environment and build a high-quality living environment. We should pay attention to the rationality of urban green space layout, improve the function and quality of green space service, enhance the openness and transparency of landscape planning and construction procedures, improve public participation and enthusiasm in green space construction, and attach significance to public satisfaction with urban greening, so as to ensure that urban construction reflects public opinion.

3. Flexible mechanism

Nowadays, cities are experiencing constant development and changes, especially in the period of rapid development of our country. So some changes should be made in the construction of ecological civilization in response to the resilient environmental changes, including the adaptation of ecosystem at different stages, meeting different object demands, as well as its adaptation to different scale ranges, different objectives and tasks. Landscape architecture needs to adopt dynamic and highly flexible planning methods as guidance, construct a comprehensive development strategic framework to cope with this increasingly dynamic change, and bring possible dynamic factors into the layout of ecological civilization construction, so as to adapt to the development and changes of urban society and economy. In order to resolve the contradiction between construction and protection, between short-term and long-term development, we should strengthen the flexibility and adaptability of urban green space so that we can achieve the sustainable development of cities.

4. Collaborative platform

The construction of ecological civilization is inseparable from the close cooperation at various professional levels, such as landscape architecture, urban planning, forestry, hydraulics, economy and transportation, and also requires the cooperation among government, enterprises and ordinary people. Urban landscape management departments should take the initiative to build a collaborative platform and an inter-agency coordination and cooperation system in order to break the administrative boundaries between various agencies. Through the joint-cooperation of various stakeholders, different agencies and specialties can provide scientific and technological supports for the construction of ecological civilization from different perspectives, so as to solve the problems in the construction of ecological civilization and jointly promote the construction of beautiful China.

4 Vision into the future

As the core of China's ecological human settlement development, landscape architecture is also inevitably an important force to push forward China's ecological civilization construction. At the beginning of the 14th Five Year Plan, which is also a new starting point for the long-term goal of 2035, landscape architecture should take the construction of beautiful China as its major responsibility. It should take the initiative to respond to national policy, enrich the theoretical basis, cultivate innovative thinking and strengthen capacity building in this industry. Besides, it should also continuously provide theoretical, method and technical support for the construction of ecological civilization in China, act as a pioneer for ecological civilization construction, and usher in a new chapter of landscape architecture to meet the arrival of a new era of socialist ecological civilization.

目 录
Contents

1 公园城市与国土生态空间
Park City and Territoral Ecological Space

公园城市理念下的森林生态系统 — 18
——四川成都市东进区森林生态系统营建研究
Forest Ecosystem under Park City
— Study on Forest Ecosystem Construction in Dongjin District, Chengdu City, Sichuan Province

京津冀协同发展下的绿色空间体系 — 24
——河北石家庄城市绿地系统规划
Green Space System under the Coordinated Development of Beijing, Tianjin and Hebei
—Urban Green Space System Planning, Shijiazhuang City, Hebei Province

构建蓝绿交织、清新明亮、水城共融的生态城市格局 — 28
——北京城市副中心生态空间系统规划研究
Constructing the Pattern of Eco-city with Blue and Green Interweaving, Fresh and Bright, and the Integration of Water and City
— Study on the Eco-space System Planning of Beijing Sub-center

可持续发展的人与自然和谐共生生活圈 — 34
——山西晋城市环城生态系统规划
Sustainable Development of Human and Nature Harmonious Symbiotic Life Circle
— Planning of the Sub-urban Ecosystem, Jincheng City, Shanxi Province

城市·遗产·风景交融互动的山地绿色空间体系 — 40
——河北承德市绿地系统规划
Mountain Green Space System of City, Heritage and Landscape Interaction
— Planning of Green Space System, Chengde City, Hebei Province

三生空间统筹，弹性发展的风景道 — 46
——内蒙古乌兰察布市草原生态风景道规划
Planning of Landscape Road of Flexible Development
— The Planning of Ecological Scenic Road of Ulanqab Grassland in Inner Mongolia

活化利用文化遗产的 4R 模式 — 52
——山西晋城市陵川县文化风景道规划
4R Model of Activating and Utilizing Cultural Heritage
— Planning of Cultural Routes in Lingchuan County, Jincheng City, Shanxi Province

2 园林博览会与事件性景观
Garden Exposition and Event Landscape

为城市绽放的花园 　　　　　　　　　　　　　　　　　　　　64
——2019年南阳世界月季洲际大会博览园
The Garden Blooming for the City
— 2019 WFRS Rose Regional Convention EXPO Park in Nanyang

棕地生态修复的城市绿色引擎 　　　　　　　　　　　　　　70
——河北省第二届园林博览会（秦皇岛）规划设计
Urban Green Engine for Brownfield Ecological Restoration
— Planning and Design of the Second Garden Expo Park (Qin Huangdao) of Hebei Province

"水韵园博、圆梦人居"的城市双修典范 　　　　　　　　　76
——第十三届中国园林博览会长沙市申办概念规划
A Model of City Betterment and Ecological Restoration Programs with "Rhyming Water Garden Expo, Living Dream Human Habitat"
— The 13th China International Garden Expo Conceptual Planning for Changsha's bid

"事件景观"新公共空间作为城市主义的催化剂 　　　　　80
——2012年第三届亚洲沙滩运动会主会场及公园规划设计
New Public Space of "Event Landscape" as the Catalyst of Urbanism
— The 3rd Asian Beach Games Main Venue and Park

3 植物园
Botanical Garden

博自然之物，绽盛世之花 　　　　　　　　　　　　　　　　92
——国家植物博览馆园区总体规划
Boasting Things of Nature, Blooming Flowers of Prosperous Times
— National Botanical Museum Park Planning

黄河流域高质量发展的生态文化新地标 　　　　　　　　　98
——中原黄河植物园总体规划
A New Ecological and Cultural Landmark of High Quality Development in the Yellow River Basin
— Planning and Design of Zhongyuan Yellow River Botanical Garden

近自然城市生物多样性保护的绿色休闲地 　　　　　　　　　　　　　　　　102
——河北石家庄植物园规划设计

A Green Leisure Place of Near-nature Urban Biodiversity Protection
—— Planning and Design of Shijiazhuang Botanical Garden in Hebei

山海城市的绿色启动器 　　　　　　　　　　　　　　　　　　　　　　　106
——山东烟台植物园规划设计

Green Starter of Mountain and Sea City
—— Planning and Design of Yantai Botanical Garden in Shandong Province

4 生态修复与郊野公园
Ecological Restoration and Country Parks

黄土台塬的生态系统再生 　　　　　　　　　　　　　　　　　　　　　　120
——山西晋中市百草坡森林植物园规划设计

Ecosystem Regeneration in Loess Tableland
—— Planning and Design of Baicaopo Forest Botanical Garden, Jinzhong City, Shanxi Province

煤矸石堆放区的绿色转型 　　　　　　　　　　　　　　　　　　　　　　130
——山西晋城白马寺沉陷区生态综合治理工程规划设计

Green Transformation of Coal Gangue Stacking Area
—— Ecological Comprehensive Improvement Project of Baimasi Subsidence Area, Jincheng City, Shanxi Province

绿色郊野空间的弹性节约 　　　　　　　　　　　　　　　　　　　　　　136
——山东烟台夹河生态郊野公园规划设计

Elastic Saving of Green Country Space
—— Planning and Design of Jiahe Ecological Country Park, Yantai City, Shandong Province

5 海绵城市与滨水空间
Sponge City and Waterfront Space

城市与自然之间的集雨绿地 　　　　　　　　　　　　　　　　　　　　　150
——河北迁安滨湖东路东侧绿带规划设计

Rainwater Harvesting Greenbelt between City and Nature
——Rainwater Collection Green Space between the City and Nature, Qian'an City, Hebei Province

延展地脉文脉的滨水空间 158
——河北迁安佛山公园规划设计
A Waterfront Landscape Design that Extends the Natural and Cultural Veins
— Planning and Design of Buddhist Culture Park, Qian'an City, Hebei Province

传承晋商文化的绿色枢纽 164
——山西晋中晋商文化公园规划设计
A Green Hub for Inheriting the Culture of Shanxi Merchants
— Planning and Design of Jin Commercial Cultural Park, Jinzhong City, Shanxi Province

炎帝文化复兴的生态文化廊道 170
——山西高平炎帝文化公园规划设计
Ecological Cultural Corridor of Yandi Cultural Renaissance
— Planning and Design of Yandi Cultural Park, Gaoping City, Shanxi Province

卫星城区催化剂 176
——山东济阳安澜湖水公园
Catalyst of Satellite District
— Anlan Lake Park, Jiyang City, Shandong Province

6 城市更新与绿色开放空间
Urban Renewal and Green Open Space

从棕地到绿色基础设施 192
——工业城市转型视角下的山西晋中铁路公园更新设计
From Brownfield to Green Infrastructure
— The Renewal Design of Railway Park from the Perspective of Industrial City Transformation, Jinzhong City, Shanxi Province

承载非物质文化遗产与生活的自然空间 200
——山西晋中社火公园规划设计
The Natural Space that Carries Intangible Cultural Heritage and Life
— Planning and Design of the Spring Festival Parade Culture Park, Jinzhong City, Shanxi Province

活力的绿色开放空间 206
——青海西宁五一路口绿地广场设计
Vigorous Green Open Space
— Wuyi Road Intersection Green Square Design, Xining City, Qinghai Province

城市棕地中铁路遗迹的重生 　　　　　　　　　　　　　　　　　　212
——青海西宁通海桥公园
Rebirth of Railway Heritage in Urban Brownfield
— Tonghaiqiao Park, Xining City, Qinghai Province

7 文化展园与艺术花园
Cultural Garden and Art Garden

基于文化遗产复兴的设计实践 　　　　　　　　　　　　　　　　226
——河南焦作七贤园
The Renaissance of Handicraft's Cultural Heritage in Landscape Architecture
— Seven Sages Garden, Jiaozuo City, Henan Province

城市地域文化的多样性表达 　　　　　　　　　　　　　　　　　234
——河南南阳五圣园
Diversified Expression of Urban Regional Culture
— Five Saints Garden, Nanyang City, Henan Province

唤醒城市记忆 　　　　　　　　　　　　　　　　　　　　　　　240
——2019 世界月季洲际大会河南邓州园
Bring Back the City's Historical Memory
— Dengzhou Garden of 2019 WFRS Rose Regional Convention, Henan Province

诗意栖居的文化生活空间 　　　　　　　　　　　　　　　　　　246
——河北保定清泽园
Cultural Life Space of Poetic Dwelling
— Qingze Garden, Baoding City, Hebei Province

后记　　　　　　　　　　　　　　　　　　　　　　　　　　　251
Postscript

1
公园城市与国土生态空间
Park City and Territoral Ecological Space

当前，在中国特色社会主义进入新时代的重要历史时期，我国城市生态和人居环境面临着新的形势和全新挑战。2018年春节前夕，习近平总书记在成都市天府新区视察时作出了"突出公园城市特点，把生态价值考虑进去"的重要指示。自此，公园城市作为一个新的城市发展理念，既对城市发展提出了新的目标指引，也对我国城市的生态和人居环境建设提出了更高要求。如何更好地适应新时代的发展，为我国公园城市建设提供正确的价值导向和科学引导，是风景园林、城乡规划等相关学科需要思考的重要议题。

公园城市的提出在城市的管理、经营思想理念方面作出了重要革新，反映了我国城市生态和人居环境建设的最新认知水平，是我国城市建设理念的历史性飞跃，也是解决当前我国城市发展问题的最佳答案。公园城市在园林城市、生态园林城市等发展模式的基础上进一步提升了生态文明建设和绿色发展的内涵和目标，其战略意义主要体现在以下几个方面。公园城市建设是中央五大发展理念的生动实践，建设公园城市不单单是建设城市公园，而是以建设和谐、宜居城市为目标的经济、社会、环境协调发展。公园城市建设突出以人民为中心的发展思想，核心在于"公"，面向公众，公平共享。通过打造人人可享受的高品质生活环境以满足人民日益增长的美好生活需求，打造开放性、可达性、亲民性的公园体系，用公园体系这一最具吸引力的公共产品打造人人向往的人居环境，提升城市综合竞争力。公园城市建设是构建人与自然和谐共生的理想蓝图，以创造优良的生态人居环境作为中心目标，将城市建设成为人与自然和谐共生的美丽家园。继承发展山水城市理念和中国园林文化，强调天人合一，提倡人工环境、自然环境和文化底蕴的协调发展，从而实现"生境、画境、意境"的高度统一。

在公园城市建设的宏观背景下，传统绿地系统规划亟待进行转型适应。首先要面向国土空间规划，规划范围从城市扩展到城乡。长久以来，中国的城市绿地系统规划作为城市总体规划的细化和深化，根据城市总体规划制定各类城市绿地的发展指标、建设工作以及市域大环境绿化的空间布局。因此，传统绿地系统规划偏重于城市建设用地内绿地的规划建设，而"其他绿地""乡村绿地"等表述长期停留在概念模糊的层面。随着2017年新版《城市绿地分类标准》的实施和2019年《城市绿地规划标准》的颁布实施，开始强调城市建设用地外绿地对市域绿地系统的重要作用，但是仍存在边界划定、管理权属等问题。因此，随着国土空间规划体系的建立，亟待规划人员从城乡一体的大区域角度出发提出科学合理的解决方案。

技术流程上，从绿地填空到适宜性布局。传统城市绿地系统规划多在城市其他建设用地布局完成后进行，规划范围主要在城市建设用地内，而其他连续的自然要素（如山体、水域等）受到行政区划的影响，并没有纳入其中，未能突破行政管理边界限制统筹考量广域、连续的生态空间体系。近年来，在建设用地外绿地的规划方面进行了大量实践和探索，在新时代背景下要继续加强城乡绿地与生态、社会、文化耦合关系的综合考量，从大的广域视角突破行政边界限制，谋求城乡绿地生态系统服务价值综合效益的最大化。

辅助决策方面，从定性到定量。近几年，城市绿地系统规划在编制过程中已开始引入景观生态学、地理学等学科理论方法。但是，编制过程中仍存在缺少基础数据等问题，并且设定绿量、分期规划目标时以计划性指标为主，导致规划中远期指标的实现力度难以提高等问题产生。因此，今后需要在国土空间基础信息平台建立的基础上，实现专项规划与总体规划的底图和资源统一，增强多学科理论体系分析技术的融合运用，提升科学技术手段对规划空间布局和指标计算的支撑力度。

在规划实施阶段，从分离到协同。城市建设用地内绿地的建设工作基本能够按照规划内容实施，但城市建设用地外生态空间通常由于涉及不同的管理部门，其规划实施率难以有效把控，且有关部门间尚未建立明确的空间协同管理机制。因此在新时代国家管理体制和空间规划体系下，需要针对以往林业、环保、水利等部门对城镇开发边界外生态空间多方分管、规划管理区域重叠、层次复杂等问题，提出管控体制统一的发展策略。

本章选择了 7 个实践项目，以森林生态系统、绿地系统、生态空间、线性风景道等多元载体，探索不同规划视角下国土生态空间规划实践方法。

成都市东进区森林生态系统营建在公园城市背景下提出"森林结构基底 + 森林镶嵌单元"的整体森林营建模式，形成格局—服务—脆弱度全方位的分析手段，以森林营建和行动计划为抓手，推动公园城市示范区绿色生态系统建设和成渝双城经济圈主轴绿色枢纽建设。

石家庄市城市绿地系统规划充分把握京津冀一体化城市发展转型、空间结构调整所带来的历史机遇，以绿地建设为表率，引导城市转型，强调自然环境与人工环境的协调发展，构建"蓝绿交织、清新明亮、城绿交融"的山水城市发展构架。

北京城市副中心生态空间系统规划研究着力建设生态、游憩与文化相融合的蓝绿空间，以生态为基础塑造点线面结合的蓝绿空间格局，以游憩为特色优化生态、生活、生产融合的蓝绿空间内容，以文化为灵魂，丰富文脉延续、遗产保护的通州蓝绿空间内涵。

山西晋城郊区生态系统规划，提出了"环城生态系统"的概念，建立了区域尺度上的青山资源开发规划框架和体系。在土地适宜性评价的基础上确定区域绿地的恢复力，提出保护和利用城市周边绿地的 CMDR 体系，实现绿地的衔接，促进城市与自然的融合，最终构建文化景观规划体系。

承德市城市绿地系统规划构建"城山交融、蓝绿交织"的山地城市绿地空间特色，传承避暑山庄古典园林的思想，充分利用承德目前现有的山水资源，打造具有承德山地城市特色和世界文化遗产底蕴的京津冀山水型公园城市。

内蒙古乌兰察布草原生态风景道在弹性规划上，通过科学的规划方法和合理的保护过程促进区域生物多样性提升，打造区域农牧交错带景观整合的载体。微观尺度通过生态敏感性分析以及廊道识别，构建点、线、面三个维度的视觉景观，促进旅游资源整合与农业转型。

陵川县文化风景道规划基于文化线路理念，充分考虑现状，提出文化线路带动下的"4R"策略，对文化遗产进行保护和利用，形成人文休闲网络，激活城市活力。

At present, in the important historical period when China's socialism has entered a new era, China's urban ecology and human settlement environment are facing new situations and challenges.

On the eve of the 2018 Spring Festival, President Xi Jinping made an important instruction to "highlight the characteristics of park cities and take ecological values into consideration" during his inspection in Tianfu New District of Chengdu. Since then, as a new concept of urban development, Park City has not only put forward new goals and guidelines for urban development, but also put forward higher requirements for the construction of urban ecology and human settlement environment in China. How to better adapt to the development in the new era and provide correct value orientation and scientific guidance for the construction of park cities in China is now an important topic for landscape architecture, urban and rural planning, and other related disciplines.

Park City has made important innovation to urban planning and management, reflecting the latest cognitive level of urban ecology and human settlement environment construction in China. It is a historic leap in the concept of urban construction in China, and the best solution to the current urban development problems in China. Based on the development models of garden city and ecological garden city, park city further enhances the connotation and goal of ecological civilization construction and green development.

Its strategic significance is mainly reflected in the following aspects. The construction of park city is the vivid practice of the five major development concepts of the central government. The construction of park city is not only the construction of urban parks, but also the coordinated development of economy, society, and environment with the goal of building a harmonious and livable city. The construction of park city highlights the idea of people-centered development, and the core is open to and shared by the public. Efforts should be made to create a desirable living environment for everyone and enhance the comprehensive competitiveness of the city by creating a high-quality living environment that everyone can enjoy to meet people's growing needs for a better life, and establishing an open, accessible and people-friendly park system, which is the most attractive public product. The construction of park city is an ideal blueprint for the harmonious co-existence of man and nature, which takes the creation of excellent ecological living environment as the central goal and turns the city into a beautiful home for the harmonious co-existence of man and nature. The concept of mountain-water city and Chinese garden culture should be inherited and developed with emphasis on the unity of nature and man as well as the coordinated development of artificial environment, natural environment, and cultural heritage, so as to achieve a high degree of unity of "habitat, painting environment and artistic conception".

Under the macro background of park city construction, the traditional green space system planning needs to be transformed and adapted urgently. First, we need to plan for territorial space, which extends from cities to rural areas. For a long time, the urban green space system planning in China has been regarded as the refinement and deepening of the overall urban planning. According to the general urban planning, the development indicators, construction work and the spatial layout of urban green space are formulated for all kinds of urban green space. Therefore, the traditional green space system planning lays emphasis on the planning and construction of green space within urban construction land, while the expressions of "other green space" and "rural green space" remain vague in concept for a long time. With the implementation of the new Classification Standard for Urban Green Space in 2017 and the promulgation and implementation of the Planning Standard for Urban Green Space in 2019, the important role of the green space outside the urban construction land for the urban green space system has been emphasized, but its boundary demarcation, management rights and other issues still exist. Therefore, with the establishment of territorial spatial planning system, it is urgent for planners to put forward scientific and reasonable solutions from the perspective of urban and rural integration.

On the technical process, from green space filling to suitability layout. Traditional urban green space system planning in urban construction land layout is often made after the completion of other urban construction land, and other continuous natural elements, such as mountains, waters, etc. affected by the administrative divisions are not included, and hence it is impossible to break through the administrative boundary to take wide-area, continuous system of ecological space into consideration. In recent years, tremendous exploration and practice have been performed in the planning of the green land outside the construction land, and in the context of a new era continuous efforts should be made to strengthen the comprehensive consideration of rural and urban green space and the coupling relationship of ecology, society and culture to break through the administrative boundaries from the perspective of large wide-area restrictions and realize the maximum comprehensive benefit of urban and rural green space ecosystem service value.

For decision-making aids, from qualitative to quantitative. In recent years, theories and methods of landscape ecology and geography have been introduced into the planning of urban green space system. However, there are still problems such as lack of basic data in the compilation process, and planning targets are mainly planned when setting green quantity and planning goals by stages, which leads to problems such as difficulty in improving the realization of medium and long-term planning targets. Therefore, based on the establishment of the basic information platform for territorial space, it is necessary to unify the base maps and resources of special planning and overall planning, enhance the integration and application of multidisciplinary theoretical system analysis technology, and strengthen the support of scientific and technological means for spatial layout planning and index calculation.

In the planning implementation phase, from separation to collaboration. The construction of green space within urban construction land can be basically implemented according to the planning content, but the ecological space outside urban construction land usually involves different administrative departments, and its planning implementation rate is difficult to effectively control, and the relevant departments have not yet established a clear spatial collaborative management mechanism. Therefore, under the national management system and spatial planning system in the new era, it is necessary to put forward a development strategy of unified management and control system, aiming at the problems such as the complex overlapping planning and management of ecological space outside the urban development boundary by forestry, environmental protection, water conservancy and other departments.

In this chapter, seven practical projects are selected to explore the practical methods of land ecological space planning from different planning perspectives by multiple carriers such as forest ecosystem, green space system, ecological space, and linear scenic path.

Chengdu Dongjin District forest ecological system construction under the background of park city puts forward the "forest structure base + forest mosaic unit" overall forest construction pattern and forms an all-round analysis method of pattern - service - fragile degree. Through forest construction and action plan, the green ecological system construction of demonstrative park city and the construction of green spindle hub of Chengdu-Chongqing economic circle are actively promoted.

Shijiazhuang city green space system planning takes the historic opportunities brought by the developmental transformation and space restructure of Beijing-Tianjin-Hebei integration into full consideration. With green space construction as an example and emphasis on the coordinated development of the natural environment and artificial environment, the city draws a developmental blueprint for a scenic city of "blue and green brightness and freshness" to guide the urban transformation.

Beijing sub-center ecological space system planning focuses on building a blue and green space of ecological, recreational, and cultural fusion. It establishes a combined blue and green space pattern of point, line and plane with an ecological basis, optimizes the contents of the blue and green space of ecology, life and production with recreational characteristics, and enriches the blue and green space connotation of cultural inheritance and extension with culture as the soul.

The concept of "ring-city ecosystem" is put forward in the suburban ecosystem planning of Jincheng, Shanxi, and the framework and system of the development and planning of mountain resources on the regional scale are established. Based on land suitability evaluation, the resilience of regional green space is determined, and the CMDR system for protecting and utilizing green space around the city is proposed, so as to realize the connection of green space, promote the integration of city and nature, and finally construct the cultural landscape planning system.

The urban green space system of Chengde city is characterized by mountainous city green space of "integration of city and mountain and interweaving of blue and green". It inherits the idea of classical garden of summer resort, makes full use of the existing landscape resources of Chengde, and creates a Beijing-Tianjin-Hebei landscape park city with characteristics of mountainous Chengde city and world cultural heritage.

Above the elastic planning red line, the Ulanqab Grassland Ecological Landscape Byway in Inner Mongolia promotes the improvement of regional biodiversity through scientific planning methods and reasonable protection processes and creates a carrier for the landscape integration of regional agricultural-pastoral ecotone. At the micro scale, through ecological sensitivity analysis and corridor identification, a visual landscape with three dimensions of point, line and plane is constructed to promote the integration of tourism resources and agricultural transformation.

Based on the concept of cultural routes, Lingchuan County Cultural Routes planning fully considers the current situation and puts forward the "4R" strategy driven by cultural routes to protect and utilize cultural heritages, form a cultural leisure network and activate the vitality of the city.

公园城市理念下的森林生态系统
——四川成都市东进区森林生态系统营建研究

Forest Ecosystem under Park City
— Study on Forest Ecosystem Construction in Dongjin District, Chengdu City, Sichuan Province

完成时间：2020 年	Time of completion: 2020
建设地点：四川省成都市东进区	Construction site: Dongjin District, Chengdu City, Sichuan Province
项目面积：3976 km²	Project area: 3976 km²
建设单位：成都市公园城市建设管理局	Construction unit: Chengdu Park City Construction Administration

获奖信息：2021 年国际风景园林师联合会亚非中东地区公园和环境类卓越奖
Awards: Award International Federation of Landscape Architects Asia-Africa Middle East Region Award (IFLA AAPME) Park and Environment Excellence Award in 2021

主要设计人员：李雄 郑曦 李方正 董丽 郝培尧 林辰松 徐昉 邵明 等
Project team: LI Xiong, ZHENG Xi, LI Fangzheng, DONG Li, HAO Peiyao, LIN Chensong, XU Fang, SHAO Ming, et al

    2019 年 12 月，北京林业大学与成都市政府签署了战略合作协议，旨在共同搭建多层次、多方位合作平台；2020 年 6 月，北京林业大学园林学院与成都市公园城市建设管理局签署公园城市建设研究和科研合作协议，以校市战略合作框架协议为基础，坚持优势互补、合作共赢的理念，携手推进公园城市建设研究和实践探索。研究课题自立项以来，经过成都市公园城市建设发展研究院与北京林业大学师生团队的通力合作，探索公园城市背景下森林生态系统营建，在推进美丽宜居公园城市重点示范区技术示范上取得了系统性成果，成为成渝双城圈的重要支撑。研究课题从东进片区圈层、城乡重点地区圈层、新城圈层三个系统层级入手，分别确立营建目标与愿景，提出"森林结构基底 + 森林镶嵌单元"的整体森林营建模式，以龙泉山、沱江、绛溪河等山水对象为重点，针对各圈层自身问题，从格局—服务—脆弱度等方面进行全方位分析，层层深入形成森林营建增补方案与七大具体行动计划，并提出相应营建指标，创新性地突破了传统生态营建中单一目式的营建方法，强调了公园城市理念下的全局性、整体性的森林系统总体营建与优化，为成都市公园城市重点示范区建设提供了有力支撑，也为《成都市实施"东进"战略总体规划》绿色生态系统建设和成渝双城经济圈主轴绿色枢纽建设的实施提供了具有引领性的扎实技术指导。

    In December 2019, Beijing Forestry University and Chengdu municipal government signed a strategic cooperation agreement, aiming to jointly build a multi-level and multi-directional cooperation platform; in June 2020, the planning team and Chengdu Park City Construction Administration signed a park city construction research and research cooperation agreement, insisting on mutual benefit and promoting Park City Construction Research and Practice exploration. Since the project was established, through the cooperation of the planning team, the research project has explored the construction of forest ecosystem under the background of Park City, and achieved innovative, leading and systematic results in promoting the technology demonstration of key demonstration areas of beautiful and livable Park City, which has become an important strategic support for the construction of Chengdu-Chongqing double city economic circle. Starting from the three aspects of Dongjin District level, urban and rural key area level and new town level, this research puts forward the overall construction mode of "structure base+unit". Focusing on Longquan Mountain, Tuojiang River, Jiangxi River and other landscape objects, analyzing the problems of each level from the pattern-service-vulnerability, result to a forest construction scheme and seven specific action plans, and put forward the construction indicators, innovatively broke through the single objective construction method in traditional ecological construction, and emphasized the overall optimization of forest system under the concept of Park City, which provided strong support for the construction of key demonstration areas of Chengdu Park City, and also provided strong support for the implementation of "going east" in Chengdu, also provides guidance for the strategic master plan of Eastern Chengdu and the construction of Chengdu-Chongqing double city circle hub.

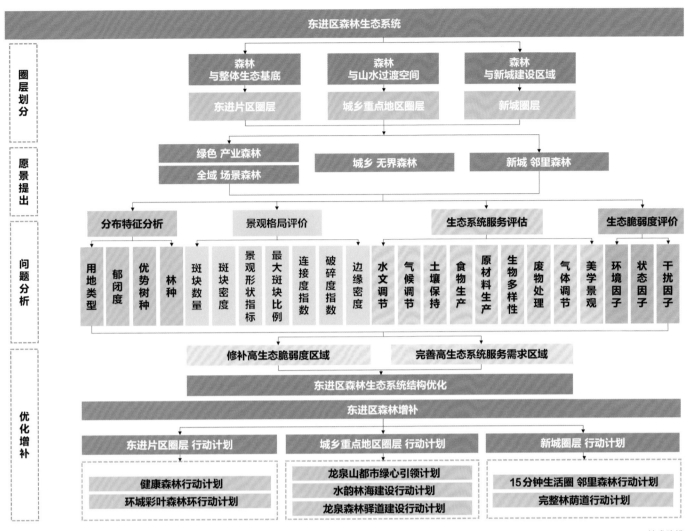

技术路线

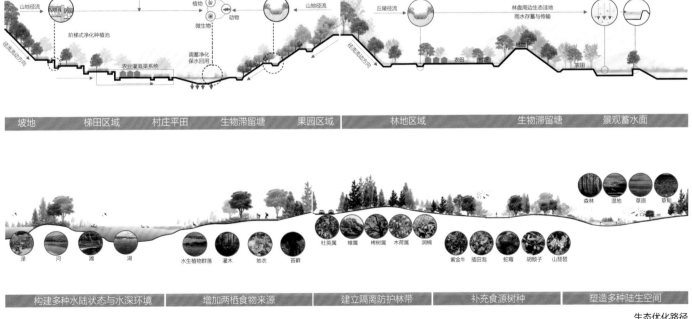

生态优化路径

1 公园城市与国土生态空间

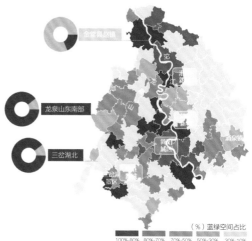

蓝绿空间占比空间分布图

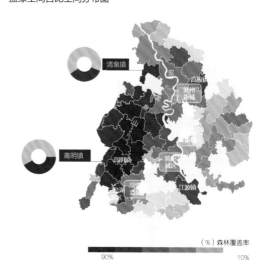

林地占比空间分布

存量

水域分布集中　林绿底色浓厚　山脉林木葱郁
蓝色空间占比　**绿色空间占比**　**森林覆盖率**
4.07%　　　　**84.82%**　　　**32.18%**

格局

总体连续局部破碎　沱江东岸斑块孤立　平原沼泽多样性高
森林连通度　　**森林破碎度**　　**森林多样性**
0.46　　　　　**0.13**　　　　　**0.73**

宜居

城郊林荫路绿视高　场地及管理满意度高　游憩设施完善
绿地空间绿视率　**绿地满意率**　　**设施完善度**
0.25　　　　　**85%**　　　　　**83%**

森林生态价值评估

功能

东进全域蓝绿空间　龙泉东麓城市绿肺
总体功能价值　　**森林气候调节服务价值**
98.60亿/年　　　**15.90亿/年**

沱江西岸保护屏障　山林湿地降温冷岛
森林水土保持服务价值　**森林气体调节服务价值**
14.70亿/年　　**11.80亿/年**

山林水系涵养源地　物质循环率待提升
森林水源涵养服务价值　**森林废物处理服务价值**
15.80亿/年　　**13.50亿/年**

山脉密林优势不足　水网支撑物种栖息
森林初级生产力服务价值　**森林生物多样性服务价值**
12.30亿/年　　**14.60亿/年**

　　规划从存量、格局、宜居、功能四大维度提出17项评价指标，对东进全域森林生态系统服务功能进行综合评估，挖掘东进森林生态价值，以期对森林功能的增补提升提供指导。

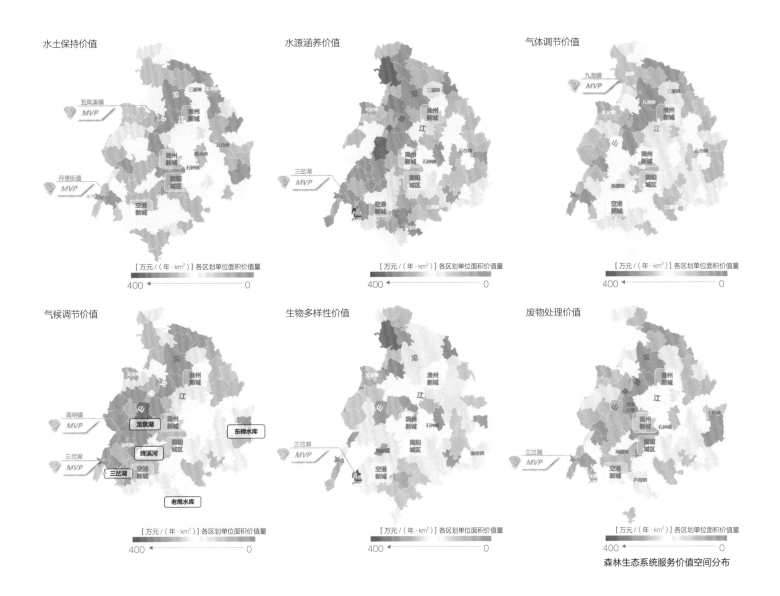

森林生态系统服务价值空间分布

规划愿景

1 公园城市与国土生态空间

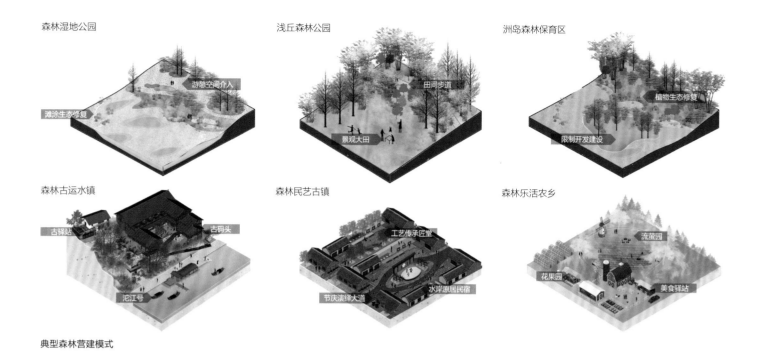

典型森林营建模式

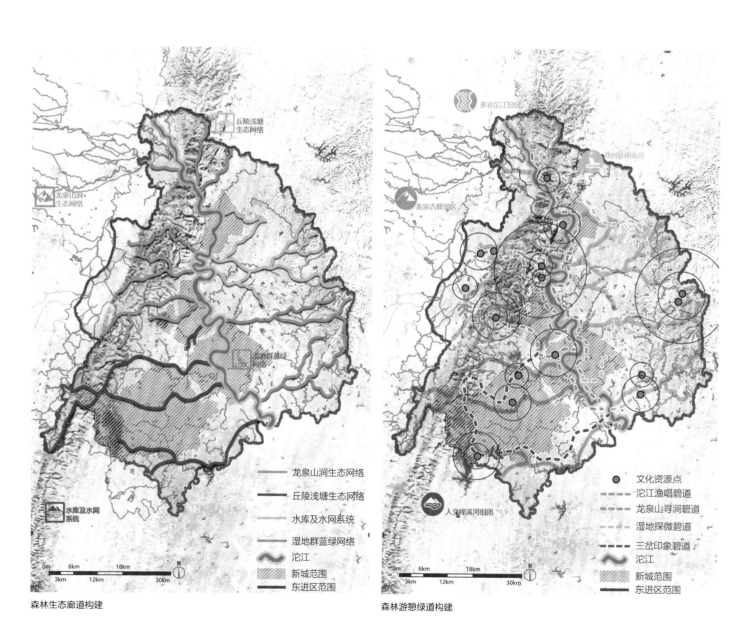

森林生态廊道构建　　　　　　　　　森林游憩绿道构建

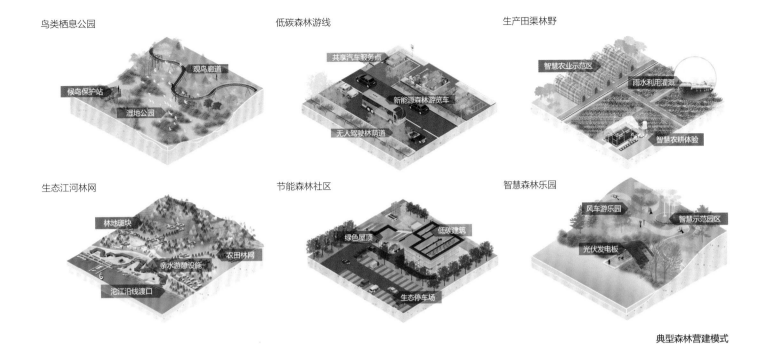

典型森林营建模式

典型营建区域：开明湖智慧生态公园

1　公园城市与国土生态空间

京津冀协同发展下的绿色空间体系
——河北石家庄城市绿地系统规划

Green Space System under the Coordinated Development of Beijing, Tianjin and Hebei
—Urban Green Space System Planning, Shijiazhuang City, Hebei Province

完成时间：2020 年
建设地点：河北省石家庄市
项目面积：14464 km²
建设单位：石家庄市园林局
主要设计人员：李 雄 张云路 马 嘉 等

Time of completion: 2020
Construction site: Shijiazhuang, Hebei Province
Project area: 14464 km²
Construction unit: Shijiazhuang Municipal Bureau of Landscape
Project team: LI Xiong, ZHANG Yunlu, MA Jia, et al

    石家庄城市绿地系统规划深入贯彻落实中央城市工作会议精神，是实施石家庄市建设绿色生态城市科学发展战略、实现"美丽幸福新石家庄"发展目标的重点规划之一。规划充分把握京津冀一体化战略背景下，城市转型发展、空间结构调整所带来的历史机遇，合理应对城市宏观层面的生态安全保障问题，主动承担石家庄作为京津冀世界级城市群中心城市所面临的绿色空间体系和人居生态环境优化挑战与机遇。规划以绿地建设为表率，引导城市转型，尽快适应京津冀协同发展战略对石家庄提出的定位和要求，协调区域经济发展与生态建设之间的关系，改善城市生态环境质量和景观效果，使石家庄城市形象得到改善，居民生活水平和质量不断提高，最终实现生态、社会、经济和谐发展的"美丽石家庄梦"。本次规划以尊重和维护自然为前提，以人民的美好生活需求为宗旨，坚持生态优先，绿色发展，实现石家庄人与人、人与自然、人与社会的和谐共生。提出城市园林建设与发展坚持以人民利益为核心，强调自然环境与人工环境的耦合协调，构建"蓝绿交织、清新明亮、城绿交融"的山水城市发展构架。

    Shijiazhuang urban green space system planning is one of the key plans to implement the spirit of the central city work conference, implement the scientific development strategy of building Shijiazhuang into a green ecological city, and realize the grand goal of "beautiful and happy new Shijiazhuang". The Plan fully grasp the Beijing-Tianjin-Hebei integration strategy background, the transformation of city development, historical opportunity brought by the spatial structure adjustment, the reasonable response to the macro level of urban ecological security problems, the initiative in Beijing-Tianjin-Hebei urban agglomeration world class center for Shijiazhuang the urban green space system facing challenges and opportunities and living ecological environment optimization. The plan take the green space construction as an example, guide the city transformation, adapt to the new requirements of the new urbanization construction for the development of Shijiazhuang as soon as possible, coordinate the relationship between regional economic development and ecological construction, improve the quality of urban ecological environment and landscape effect, improve the image of Shijiazhuang City, improve the living standard and quality of residents, and finally realize the harmonious development of ecology, society and economy "Beautiful Shijiazhuang dream". The plan takes respecting and maintaining the nature as the premise, takes the people's needs for a better life as the purpose, adheres to ecological priority and green development, and realizes the harmonious coexistence of people and nature, people and people, people and society in Shijiazhuang. The construction and development of urban landscape architecture adhere to the people's interests as the core, emphasize the coordinated development of natural environment and artificial environment, and build a landscape city development framework of "blue and green interweave, fresh and bright, city green blend".

石家庄中心城区绿线规划

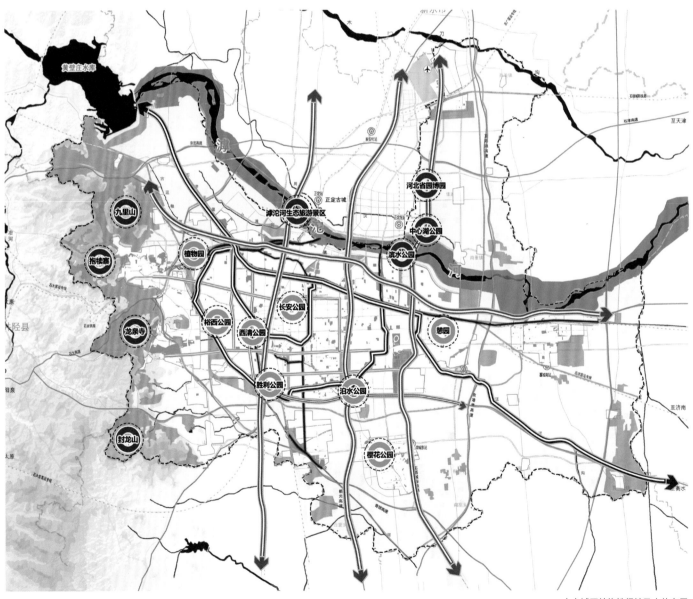

中心城区结构性绿地及水体布局

1 公园城市与国土生态空间 25

生态支撑型绿色空间

游憩服务型绿色空间

生产型绿色空间

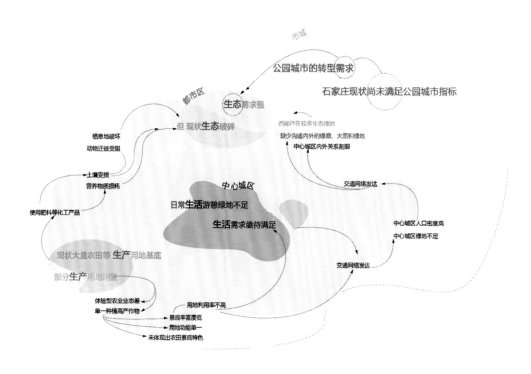
问题分析

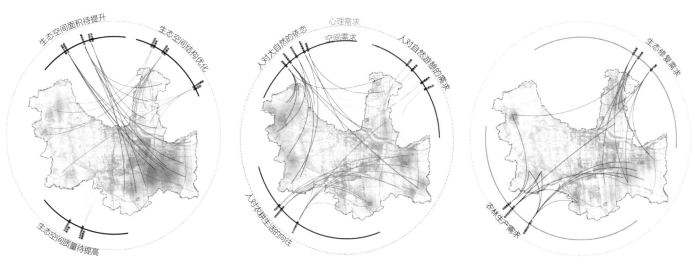

现状山水林田湖占比	现状 MSPA 分析	现状生态敏感性分析	风景名胜区	都市农业园	现状农田基底分析	现状生态敏感性分析
生态空间面积提升	生态空间结构优化	生态空间质量提高	心理需求—自然依恋	心理需求—农耕体验	农林生产需求	生态修复需求

需求分析

技术路线

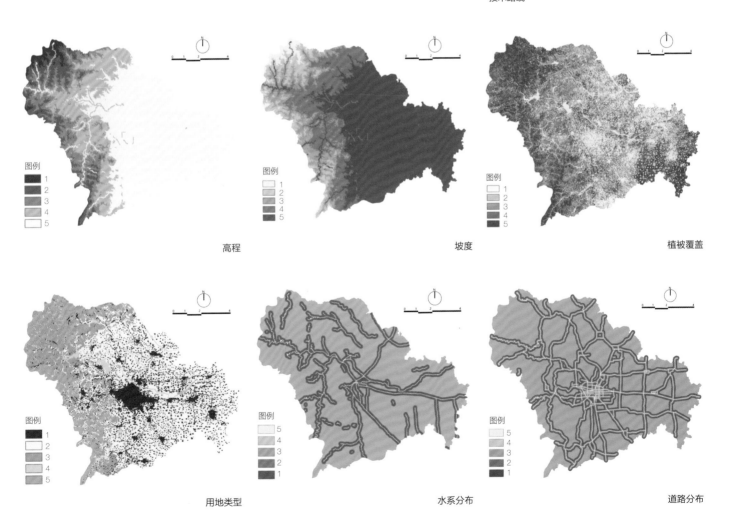

高程　　　　　　　　　坡度　　　　　　　　　植被覆盖

用地类型　　　　　　　水系分布　　　　　　　道路分布

构建蓝绿交织、清新明亮、水城共融的生态城市格局
——北京城市副中心生态空间系统规划研究

Constructing the Pattern of Eco-city with Blue and Green Interweaving, Fresh and Bright, and the Integration of Water and City
— Study on the Eco-space System Planning of Beijing Sub-center

完成时间：2017 年
建设地点：北京市通州区
项目面积：906 km²
建设单位：北京市园林绿化局
主要设计人员：李 雄 张云路 胡 楠 等

Time of completion: 2017
Construction site: Tongzhou District, Beijing
Project area: 906 km²
Construction unit: Beijing Landscaping Bureau
Project team: LI Xiong, ZHANG Yunlu, HU Nan, et al

　　本次规划研究立足实际，从本地资源出发，坚守生态原则、人本原则、文化原则、精品原则，统筹考虑建设发展与自然资源、人居环境的和谐共生，营造人景交融、文化底蕴浓厚、舒适优美的游憩环境。本研究在规划理论和规划方法上，提出与通州区资源条件和现状问题相适应的一套理论及方法，以满足北京城市副中心的发展需求。突破过去绿地规划中的"填空"不足，形成适合通州实际条件的区域蓝绿空间布局方法，为提高规划的科学性和合理性提供技术支撑。因此，本研究致力于优化通州区域组团式布局的绿地布局，完善区域生态安全格局，构建结构稳定且功能完善的生态基础空间。突出通州区域多河富水、绿色田园的特点，构建城绿交融、蓝绿交织的蓝绿游憩体系。同时着力建设生态、游憩与文化相融合的蓝绿空间，推动区域绿色基础设施网络结构的全面建设，促进生态、社会与经济的综合发展，以生态为基础塑造点、线、面结合的蓝绿空间格局，以游憩为特色优化生态、生活、生产融合的蓝绿空间内容，以文化为灵魂，丰富文脉延续、遗产保护的通州蓝绿空间内涵。

　　Based on the reality, this planning study adheres to the ecological principle, human-oriented principle, cultural principle and fine-quality principle from local resources. It considers the harmonious coexistence of construction and development and natural resources, harmonious coexistence of human settlements environment, and creates a harmonious and beautiful environment with rich cultural background and comfortable and beautiful environment. In this study, a set of theories and methods are proposed to meet the development needs of Beijing sub-center in terms of planning theory and planning methods. Break through the past green space planning "fill in the empty" insufficient, form the regional blue and green space layout method suitable for the actual conditions of Tongzhou, and provide technical support for improving the scientific and reasonable planning. Therefore, this study is devoted to optimizing the green space layout of Tongzhou District group layout, improving the regional ecological security pattern, building a stable structure and improving the ecological basic space. The characteristics of multi-river rich and green countryside in Tongzhou area are highlighted, and the green and blue recreation landscape is constructed. Meanwhile, we should focus on the construction of the blue and green space integrating ecology, recreation and culture, promote the comprehensive construction of the network structure of regional green infrastructure, promote the comprehensive development of ecology, society and economy, shape the blue and green space pattern combining the point and line with ecology as the basis, and optimize the blue and green space content of ecological life production integration with recreation as the special color. With culture as the soul, it enriches the connotation of Tongzhou blue and green space, which is the continuation of cultural context and heritage protection.

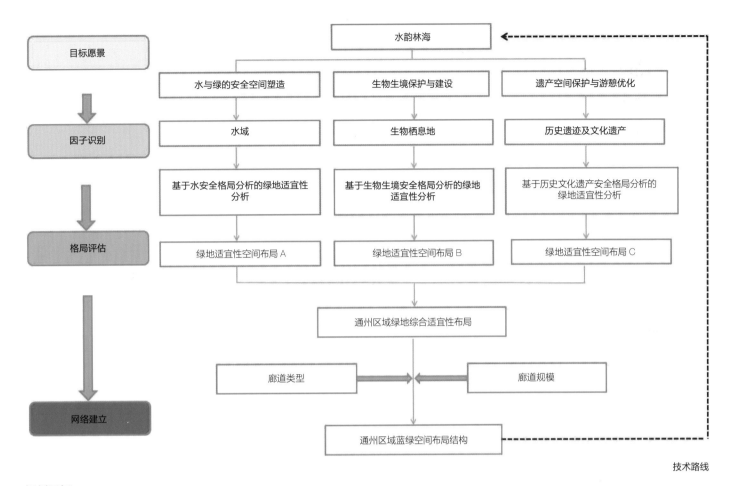

技术路线

规划目标

充分结合通州自然状况，将蓝色水网体系和绿色林网体系相互交织，使生态效益最大化。实现整体规划、结构连续的空间共同体。其次，打造三生共融的蓝绿空间格局。塑造集自然休闲旅游、绿色郊野体验、近郊游憩观光于一体的空间格局，依托良好的自然生态环境，开展绿色田园农业体验，丰富户外娱乐活动。

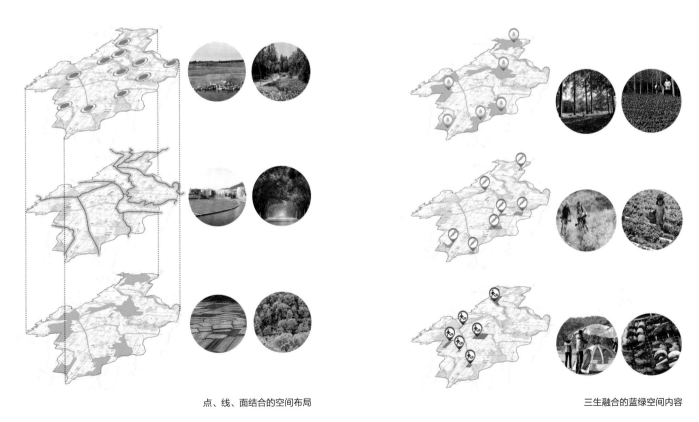

点、线、面结合的空间布局　　　　　　　　　　　三生融合的蓝绿空间内容

规划策略

强化水网连续

针对目前水系分布密集的现状以及与区域生态景观存在某些断裂的问题，通州亟待维护和强化自身贯穿区域的完整水系网络，强化水网、景观在整个城市片区的穿插和溶解，延续水网格局的连续性。

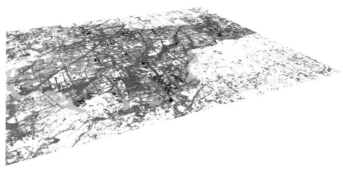

绿色空间整合

通州有密集的河网及一定面积的防护林和农田，将它们融为一体，以绿色空间的形态渗透到城市并与城郊自然景观基质融合，成为包围整个城市片区景观的绿色基质和区域绿色基础设施。

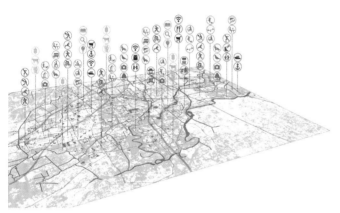

构建旅游资源连续性

构建通州河网历史文化遗产资源保护系统，增强历史文化遗产资源与河道绿网联系。构建游憩系统，增强历史文化遗产资源的可达性，提高城市整体历史文化氛围。

构建文化景观的连续性

充分发掘保护现有的河网历史资源，增强文化空间的可达性，提升城市文化的传播力，扩大公众认知力度，强化文化景观的本土特性。

重塑生境网络

从生物的栖息生境特点出发，选择远离交通干道、大型居民点和沿水近绿等空间，构建具有生物多样性的生境网络系统。

建立河网绿道慢行系统

以河网为骨架构建慢行系统，利用沿线绿道串联片区，同时加强与周围地块的功能衔接。

基于水安全格局分析的绿地适宜性分析

在对水安全进行综合评价后，将现状水系分布图、洪涝淹没风险分析图以及地表径流分析图进行叠加，可得到发展基础设施维度下的水安全格局，即基于水安全评价的绿地适宜性布局。

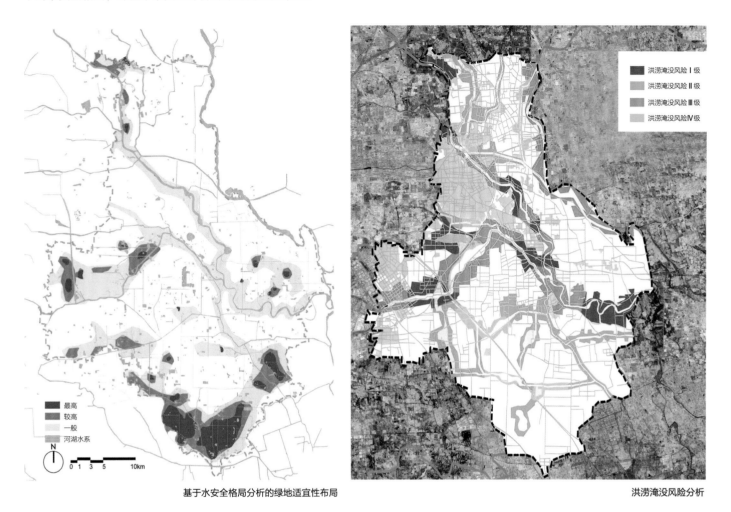

基于水安全格局分析的绿地适宜性布局　　　　　　　　　　　　洪涝淹没风险分析

基于生物生境安全格局分析的绿地适宜性分析

根据通州所在区域的生境特征，选择喜欢河流、农田和林地的大白鹭作为生物保护的指示物种，更能够反映当地的生境特征，具有较强的代表性和指示性。从生物的栖息生境特点出发，选择远离交通干道、大型居民点和沿水近绿的空间为栖息地，构建生物栖息地适宜性分级图。

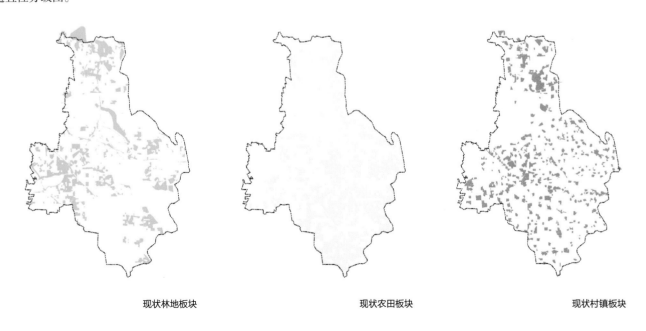

现状林地板块　　　　　　　　　　现状农田板块　　　　　　　　　　现状村镇板块

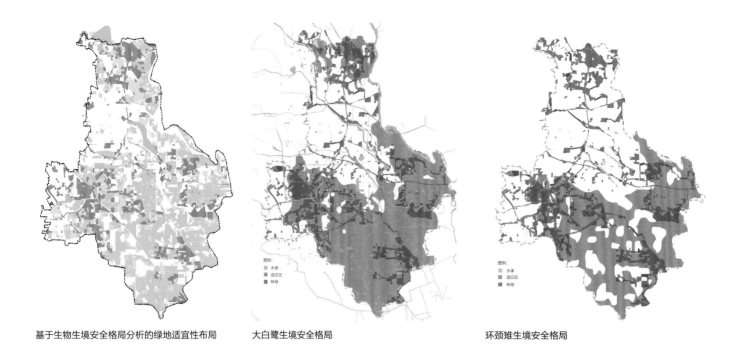

基于生物生境安全格局分析的绿地适宜性布局　　大白鹭生境安全格局　　环颈雉生境安全格局

基于历史文化遗产安全格局分析的绿地适宜性分析

通过文化遗产安全格局阻力值的分析，进行区域历史文化遗产安全格局的模拟，理想的历史文化遗产安全格局通过绿色空间串联着区域主要的遗产节点，构成了区域最具有保护价值和地域特征的空间场所。它是文化游憩的核心空间，也构成了基于历史文化遗产安全格局分析的绿地适宜性布局。

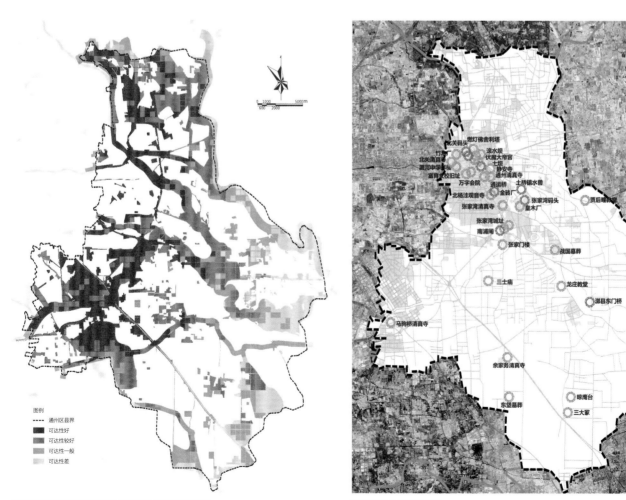

基于历史文化遗产安全格局分析的绿地适宜性布局　　现状历史文化遗产分布

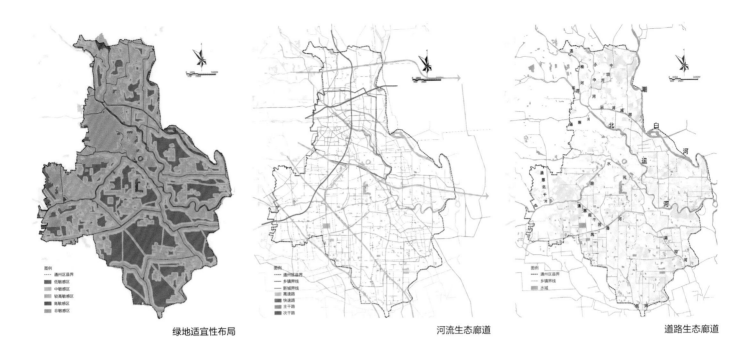

绿地适宜性布局　　　　　河流生态廊道　　　　　道路生态廊道

规划结构图

通州区域蓝绿整体空间布局呈现出"两环融城，四廊串联，绿网纵横，蓝绿交织"的结构，"两环"指环906km² 通州区域外围绿色森林带和环155km² 副中心区域绿色森林带；"四廊"包括潮白河、北运河、凉水河三条城市水系沿线绿带形成的滨水绿廊以及由六环路生态绿带形成连通区域的道路绿廊；"绿网纵横，蓝绿交织"则指通州区通过城市道路绿廊、水系廊道、公园绿地、郊野公园交织形成蓝绿生态网络。

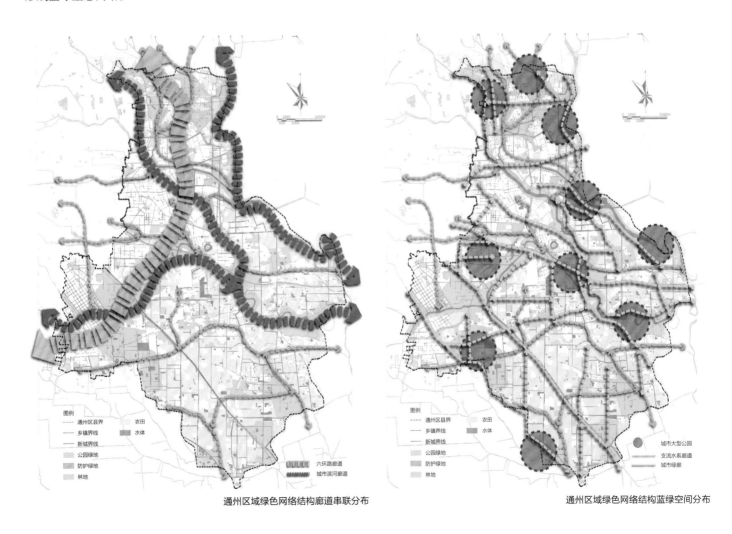

通州区域绿色网络结构廊道串联分布　　　　　通州区域绿色网络结构蓝绿空间分布

可持续发展的人与自然和谐共生生活圈
——山西晋城市环城生态系统规划

Sustainable Development of Human and Nature Harmonious Symbiotic Life Circle
—— Planning of the Sub-urban Ecosystem, Jincheng City, Shanxi Province

完成时间：2015 年	Time of completion: 2015
建设地点：山西省晋城市	Construction site: Jincheng, Shanxi Province
项目面积：396 km²	Project area: 396 km²
建设单位：晋城市规划局	Construction unit: Jincheng Planning Bureau

获奖信息： 2018 年国际风景园林师联合会亚非中东地区分析与总体规划类荣誉奖
第三届中国风景园林学会优秀风景园林规划设计三等奖
Awards: Award International Federation of Landscape Architects Asia-Africa Middle East Region Award (IFLA AAPME) Analysis and Master Planning Honor Award in 2018
The third prize of Excellent Landscape Architecture Planning and Design of the third CHSLA

主要设计人员： 李 雄 张云路 李运远 郑 曦 冯 潇 姚 朋 戈晓宇 肖 遥 林辰松 李方正 等
Project team: LI Xiong, ZHANG Yunlu, Li Yunyuan, ZHENG Xi, FENG Xiao, YAO Peng, GE Xiaoyu, XIAO Yao, LIN Chensong, LI Fangzheng, et al

　　作为一个城市重要的绿色基础设施，晋城市周边山区的研究和规划一直备受关注，但以往的规划和研究的侧重点不同、各自为政，对周边山区绿地系统的发展却鲜有整体论述。本规划立足于绿地发展的基本问题，以周边山区为研究对象，提出了"环城生态系统"的概念。该生态系统的定位是构建一个人与自然和谐的生命循环。对其生态功能、休闲功能和文化功能进行较为深入系统的研究，建立了区域尺度上的青山资源开发规划框架和体系。

　　规划基于"弹性规划"的概念，在土地适宜性评价的基础上确定区域绿地的恢复力，提出保护和利用城市周边绿地的 CMDR 体系。针对不同的弹性空间提出了四种不同的运作模式，即保护、改造、发展和振兴。这四种模式用于划分区域绿地，控制和引导开发建设内容。构建"环城生态系统"的绿道系统，实现绿地的衔接，促进城市与自然的融合，最终构建文化景观规划体系，塑造东方"山水"园林城市形象，使景观规划成为文化的保护者和传承者。

　　As a city's important green infrastructure, the research and planning of the surrounding mountain areas in Jincheng city have been closely regarded, but past planning and research focus on different and separate aspects, and the development of the green system in surrounding mountain areas has rarely been talked as a whole. This study is based on the fundamental problems of the development of green land, and set the surrounding mountain areas as the research object, propose the concept of "circular loop city ecosystem". This ecosystem is positioned to construct a life loop for human and natural harmony. Its ecological function, leisure function and culture function were studied in a relatively deep and systematic way, and that framework and system for the development planning of mountain resource on this regional scale was established.

　　Study is based on the "resilience planning" concept. Firstly, determining the resilience of green spaces of the region based on the evaluation of the land suitability, and putting forward the "CMDR" system to protect and utilize the green space of the sub-urban city. Four different modes of operation are proposed for different resilient spaces, which are conservation, modification, development, and revitalization. The four modes are used to divide green spaces of the region, control and guide the development and construction content. Moreover, they further construct that greenway system of the "circular loop city ecosystem", realize the connection of the green spaces, promote the integration of city and nature, and finally construct the heritage landscape planning system, and shape the site with the image of the east "shanshui" garden city, make the landscape planning become the protector and the inheritor of the culture.

现状图

现存问题

城市发展的需求与土地资源之间存在着强烈的冲突，大量的天然山地、水体和林地被城市开发建设所占据；项目基址西邻玉屏山，采石场和采矿业造成的破坏严重，导致植被覆盖率和生态环境较差；由于场地被划分为不同的行政区，每个区域都分别规划和建造，因此没有在整体层面上进行思考和规划，在规划中很多文化遗产景观缺乏保护和管理，脱离了总体规划和建设要求。

CMDR规划系统示意

主要理论

"CMDR"规划系统：基于土地适宜性评估，提出保护和利用城市周围绿地的指导方针。绿地与娱乐活动紧密相连，特定的绿地与特定的娱乐活动相对应。因此，景观设计应考虑娱乐活动对生态景观娱乐功能的影响。根据生态和景观敏感性之间的差异，并与其他外部标记结合，评估适宜性安全模式，对不同的适宜性进行分类，并制定有针对性的开发策略。

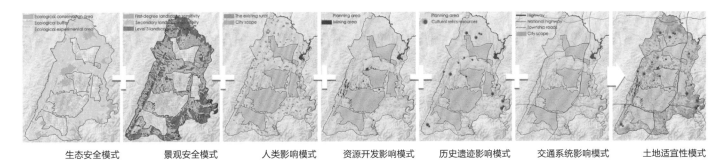

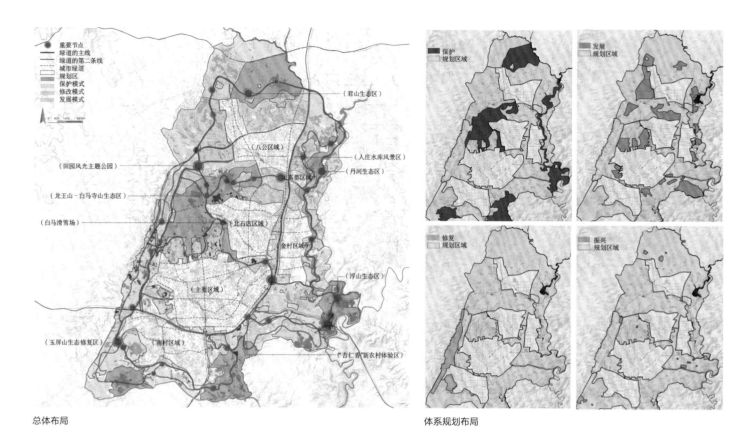

总体布局　　　　　　　　　　　　　　　　　　　　　　　　　体系规划布局

总体布局

晋城生态系统规划包括绿环、绿色走廊和多点分布三层结构，由晋城中心城区周边生态系统圈形成的绿色基础设施，可以有效地将山林、农田和城镇基地相结合，形成绿色包围城市的生态结构。根据土地对圈状绿地的适宜性评估，"CMDR"已用于将整个"环城生态系统"映射成八个功能区，分别是北山地区（朱山）、白马寺山区、玉屏山地区、金浦山地区、白水河地区、丹江东山区、龙王山—丹江走廊和白马寺山—丹江走廊。

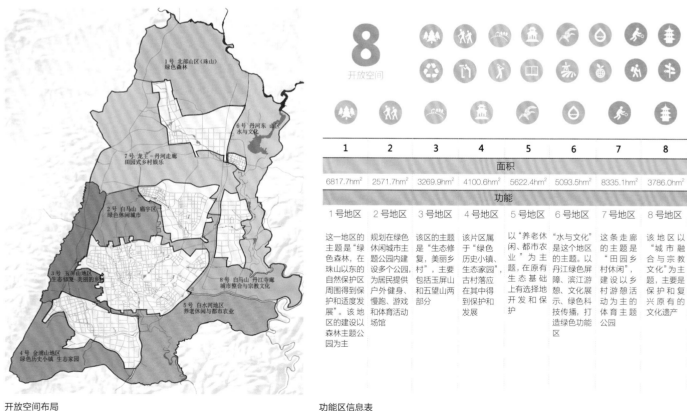

开放空间布局　　　　　　　　　　　　　　　　　　　　　　　　功能区信息表

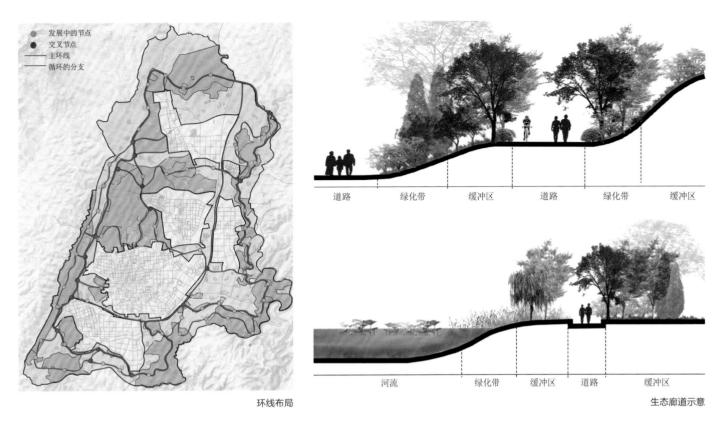

环线布局　　　　　　　　　　　　　　　　　　　　　　　　　　　生态廊道示意

布局原则

考虑整个城市的未来，使生态系统的绿色道路成为城市绿色道路的重要组成部分，与其他绿色道路连接起来；充分开发生态系统内的自然生态和历史人文资源，利用网络结构的绿色通道形式将其整合成完整的体系。因地制宜，依托景观和道路系统，建设高度可操作的绿色通道体系。

田园风光效果

1　公园城市与国土生态空间　　37

村庄慢行交通剖面

乡村休闲型慢行交通剖面

生态型慢行交通剖面

交通体系规划

　　村庄慢行交通用于文化展示、休闲观光和体育健身活动，营造绿色、安静的空间，对农村居民开放，提高人们的生活质量，提升周边地区土地价值，增强乡镇活力和吸引力。乡村休闲型慢行交通连接森林公园、生态旅游区、风景旅游区，为体育赛事、农业体验、节日习俗和乡村美食展示提供了场所。

八景图

矿区修复效果

工业遗留改造效果

1 公园城市与国土生态空间

城市·遗产·风景交融互动的山地绿色空间体系
——河北承德市绿地系统规划

Mountain Green Space System of City, Heritage and Landscape Interaction
— Planning of Green Space System, Chengde City, Hebei Province

完成时间：2019 年	Time of completion: 2019
建设地点：河北省承德市	Construction site: Chengde City, Hebei Province
项目面积：130 km²	Project area: 130 km²
建设单位：承德市园林管理中心	Construction unit: Chengde Garden Management Center
获 奖 信 息：2019 年国际风景园林师联合会亚非中东地区分析与总体规划类荣誉奖	Awards: Award International Federation of Landscape Architects Asia-Africa Middle East Region Award (IFLA AAPME) Analysis and Master Planning Honor Award in 2019
主要设计人员：李 雄 张云路 马 嘉 李 正 郝培尧 林辰松 徐 桐 肖 遥 邵 明 等	Project team: LI Xiong, ZHANG Yunlu, MA Jia, LI Zheng, HAO Peiyao, LIN Chensong, XU Tong, XIAO Yao, SHAO Ming, et al

承德，中国著名的历史文化名城，拥有最大的中国皇家离宫避暑山庄和外八庙世界文化遗产，在世界上享有盛誉。承德作为典型的山地城市，适宜开发利用的平地较少，建设用地资源紧张。而城市周边的浅山区域正面临强烈的开发建设需求。突破"建设用地中做绿地系统规划"的传统逻辑，统筹规划建设用地内外的绿色空间，充分利用城区内部及周边的山水人文资源，构建"城山交融、蓝绿交织"的山地城市绿色空间体系。

同时传承避暑山庄古典园林的思想，践行生态文明理念，坚持生态优先、绿色发展，充分利用承德的山水资源，严格落实城市外围生态控制区域，实现城市内部组团间、城市内部与外部自然环境的沟通联系。以人民为中心，营造优美、安全、舒适、共享的城市公共空间。构建蓝绿交织、城山共融、多组团集约发展的生态城市布局，创造优良人居环境，实现人与自然和谐共生，建设美丽家园。提升城市绿地空间品质与文化品位，打造具有承德山地城市特色和世界文化遗产底蕴的京津冀山水型公园城市。

Chengde, a famous historical and cultural city in China, has the largest Chinese royal summer resort and eight temples outside which are the world cultural heritage, enjoying a high reputation in the world. As a typical mountainous city, Chengde has less flat land suitable for development and utilization, and the resources of construction land are limited. However, the shallow mountain area around the city is facing a strong demand for development and construction. Breaking away from the traditional logic of "green space system planning in the construction land", overall planning of the green space inside and outside the construction land, making full use of the landscape and cultural resources inside and around the city, to build a mountain city green space system of "blending city and mountain, blue and green".

Inheriting the thought of the classical garden of the Mountain Resort, practicing the concept of ecological civilization, adhering to the ecological priority and green development, use of the landscape resources of Chengde, strictly implementing the ecological control area outside the city, and realizing the communication and connection between urban groups, the city and the external natural environment. We will create a beautiful, safe, comfortable and shared urban public space with people as the center. We will build an eco-city layout of blue and green interweaving, urban and mountain integration, and multi-group intensive and compact development, create a fine living environment, achieve harmonious coexistence between man and nature, and build a beautiful home. We will improve the quality of urban green space and cultural taste, and build a Beijing-Tianjin-Hebei landscape park city with the characteristics of Chengde mountain city and the world cultural heritage.

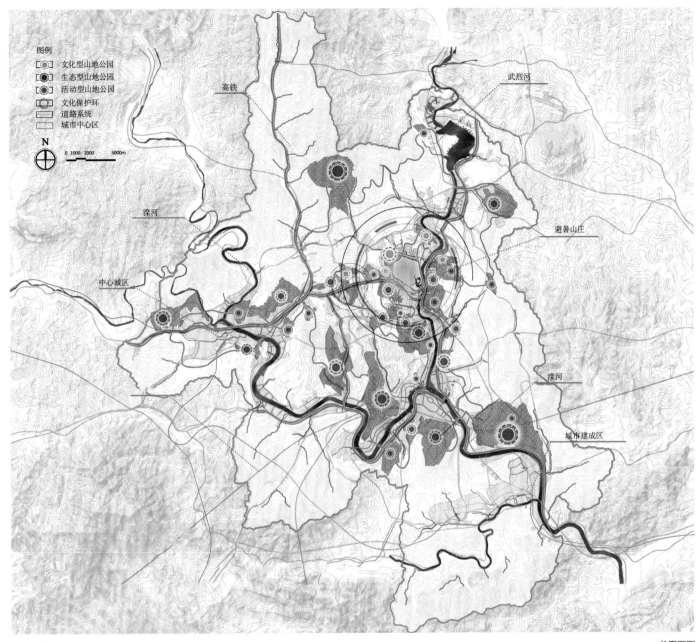

总平面图

总体规划

依托避暑山庄、外八庙、其他现存或损毁文化遗产及缓冲空间，形成点状分布的城市文化遗产空间集群。利用城市内部和周边的自然山水资源和生态空间，形成生态环境与城市空间交融渗透的绿色基底。以绿道串联城市内外各类观赏、游憩、体验的绿色空间。

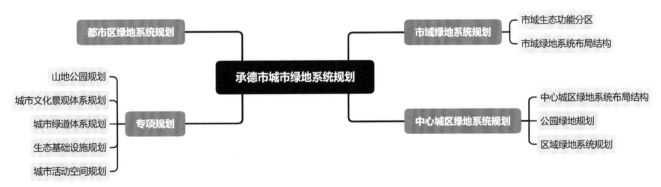

技术路线分析

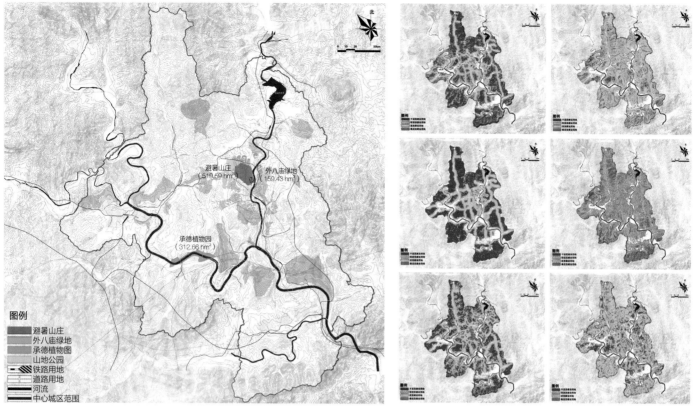

中心城区区域绿地结构　　　　　　　　　　　　　　　　　　　　　山地公园评价分析

中心城区绿地系统结构

　　一环融城、一心映城、两脉润城、多廊串联、多园均布。一环：以中心城区外围自然山体为依托形成的环城生态绿环。一心：依托被城市组团环绕的山地森林资源，形成强化城市中心生态功能的生态绿心。两脉：以滦河及武烈河为依托构建的城市生态景观绿脉。多廊：依托城市主干道及河流水系，构建城市绿廊。多园：指城市公园绿地、山地公园等。

城市内部绿地景观效果

42　理地营境　生态文明建设背景下风景园林实践

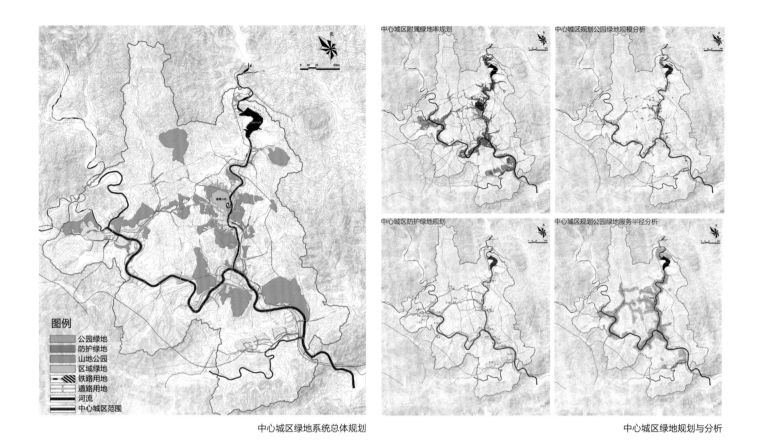

中心城区绿地系统总体规划　　　　　　　　　　　　　　　　　　中心城区绿地规划与分析

中心城区绿地系统规划

在中心城区绿地系统及专项规划层面，引领城市绿地从单一绿色空间扩展到绿色基础设施，科学引导"城山交融、蓝绿交织"城市绿地系统格局建立。以各类型绿地为载体，开展了防灾避险、雨洪管理、植物景观、绿道体系等专项规划，以满足居民对绿色福祉的多元化需求。

城市周边山地景观效果

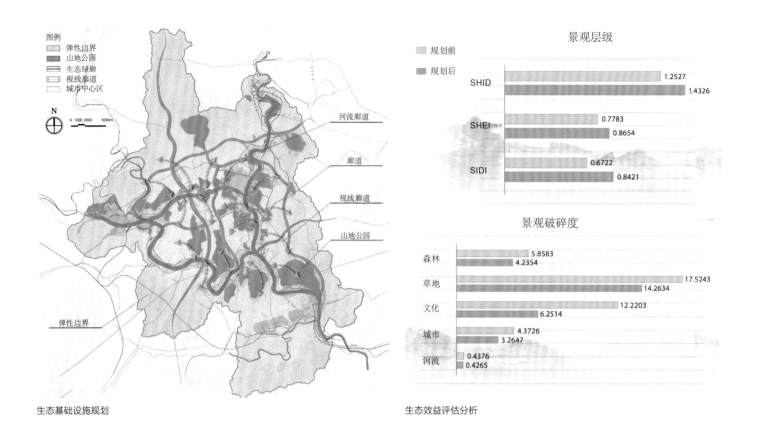

生态基础设施规划

生态效益评估分析

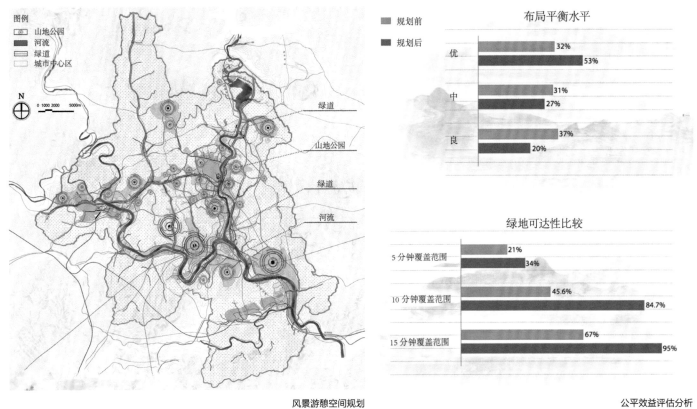

风景游憩空间规划　　　　　　　　　　　　　　　　　　公平效益评估分析

中心城区鸟瞰

1　公园城市与国土生态空间　45

三生空间统筹，弹性发展的风景道
——内蒙古乌兰察布市草原生态风景道规划

Planning of Landscape Road of Flexible Development
—— The Planning of Ecological Scenic Road of Ulanqab Grassland in Inner Mongolia

完成时间：2017 年	Time of completion: 2017
建设地点：内蒙古乌兰察布	Construction site: Ulanqab (Inner Mongolia Autonomous Region)
项目面积：1368 km²	Project area: 1368 km²
建设单位：乌兰察布市规划局	Construction unit: Ulanqab Municipal Planning Bureau

获 奖 信 息：2018 年国际风景园林师联合会亚非中东地区分析与总体规划类荣誉奖
Awards: Award International Federation of Landscape Architects Asia-Africa Middle East Region Award (IFLA AAPME) Analysis and Master Planning Honor Award in 2018

主要设计人员：李　雄　姚　朋　戈晓宇　李方正　王　鑫　等
Project team: LI Xiong, YAO Peng, GE Xiaoyu, LI Fangzheng, WANG Xin, et al

　　该项目所在的乌兰察布（内蒙古自治区）位于中国北方农牧交错带。它面临着气候变化、地理特征、民族文化和经济发展的复杂压力，借助风景道规划的契机，该项目在区域范围内形成了一个富有弹性的景观框架。该项目梳理了其穿越的农牧交错带的风景道目前面临的挑战，并在弹性规划层面提出了全面而丰富的区域景观规划策略。

　　从宏观角度来看，该项目为未来区域的生态、景观和人文保护与发展提供了明确的规划指导。同时，本项目通过透彻的数据分析和科学的规划方法，通过合理的保护过程促进区域生物多样性的提高，成为区域农牧交错带景观整合的载体。此外该项目还着眼于协调规划框架与当地生产生活，促进区域经济转型。弹性规划框架为广大从业者和学生去理解区域范围内的规划、分析和实施过程提供了一个很好的范例。

　　对于微观尺度的规划设计，主要从生态层次、景观层次和产业层级三个方面对区域进行了规划。分别通过生态层次的敏感性分析以及生态廊道的识别、景观层次的对于点、线、面视觉景观的整合，以及产业层次的对于旅游资源的整合与农业转型，对该区域进行了针对性的规划。

Ulanqab (Inner Mongolia Autonomous Region), where the project is located, is in the agro-pastoral ecotone area in northern China. It is facing the complex pressures of climate change, geographical features, ethnic culture, and economic development. With the opportunity of scenic byway planning, the project formed a resilient landscape framework at a regional scale. The project sorted out the current challenges in the agro-pastoral ecotone area that the scenic byway traverse, together with proposing a comprehensive and rich regional landscape planning strategy in the resilient planning red line.

This project provided clear planning guidelines for protection and development of ecology, landscape, and humanity. By thorough data analysis and scientific planning methods. The regional biodiversity will be enhanced by a rational protection process. Scenic byways will become the media of the landscape integration in the regional agro-pastoral ecotone area. In addition, the project focused on the suitability between planning framework and local production and living, which had led to the regional economic transformation. The resilient planning framework provides a good example for practitioners and students to understand the planning, analysing, and implementation process at a regional scale.

As for the micro-scale planning and design scheme, it mainly plans the region from three aspects: ecological level, landscape level and industrial level. Through the sensitivity analysis of the ecological level, the identification of the ecological corridor, the integration of the visual landscape of the point, line and surface at the landscape level, and the integration of tourism resources and agricultural transformation at the industrial level, targeted planning has been carried out for the region.

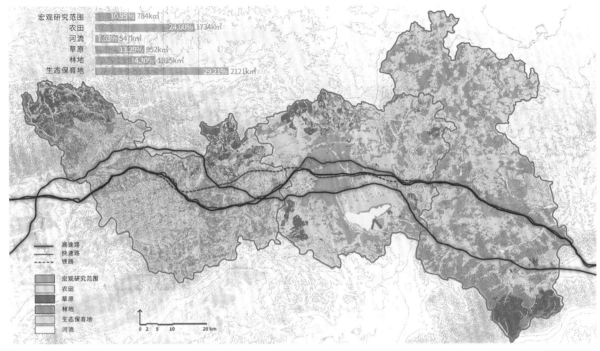

土地利用分析

本项目规划区域包括铁路1条、高速公路2条、快速路1条，全长177km。研究范围覆盖3个行政区，横跨6个自然类型，项目总面积1368km²。

当地政府在西部地区建设大量的公路、铁路等交通基础设施，穿越了山脉、水系、森林、草原等土地类型以及干旱、半干旱等不同的气候类型。然而，基础设施建设也给当地区域生态环境、地理特征、当地生产和民生带来了很大的压力和挑战。

规划区域主要集中在乌兰察布境内横跨卓子山县、济宁区、兴和县的四条交通要道，其中，G7（京乌高速）、G6（京拉高速或京藏高速）、京沪高铁和110国道东西向穿越乌兰察布。这条177km长的交通公路辐射着周围147个村庄。它们贯穿山脉、草原、湿地、城镇、农田和林地，自然景观丰富，栖息地多样。规划红线划定实施韧性战略，宏观研究面积约1368km²。

1　公园城市与国土生态空间

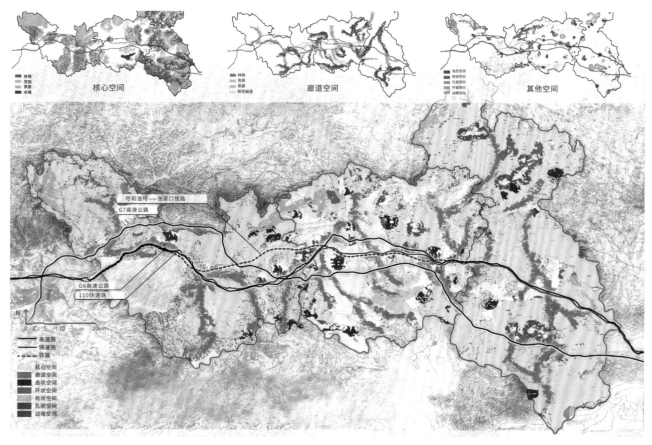

基于MSPA的生态廊道识别分析

本项目结合土地现状，应用形态空间格局分析，提取识别核心区、连接桥、孤岛、边缘、穿孔、环、分支7个生态网络要素。在被交通运输线割裂的生态地区重建生态廊道，增强生态空间网络格局的完整性，同时保护当地的生物多样性。

对气候、植被、自然化程度，以及土壤水文等因素进行叠加分析，将区域自然系统划分为最高敏感区、高敏感区、中敏感区、低敏感区和最低敏感区5个等级，并进一步叠加当地的交通基础设施路线，为区域景观的整合、保护和恢复提供了依据。

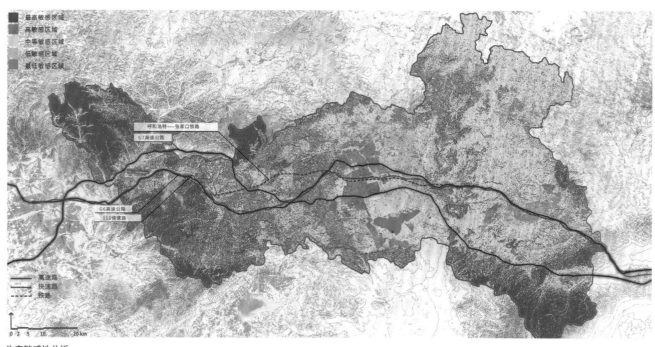

生态敏感性分析

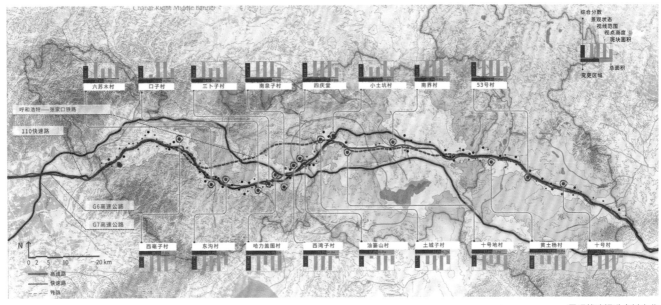

景观战略提升乡村产业

在充分研究风景道周边村民的生活特点和生产状况的基础上,通过整合景观资源,重振乡村活力,促进区域经济转型。随着与村民的沟通以及项目的推进,项目选择重点村,通过绿道建设,建立区域经济振兴框架,以此促进第一产业向第三产业的转变。发展区域经济,实现政府、企业、村民三方共赢。在确定生态廊道的基础上,考虑现有的道路和水资源的现状,规划创建了草原、农田和滨水三条绿道,利用现有收费站和服务区,在农牧交错带内增加设置驿站,形成了完善的绿道网络体系。

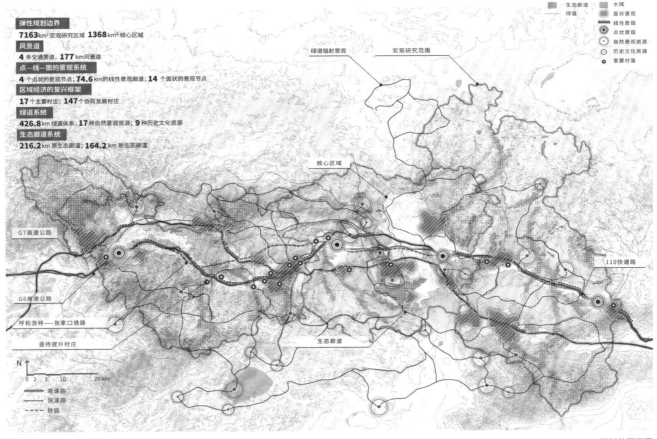

规划总平面图

1 公园城市与国土生态空间　49

风景道沿线的牧场景观

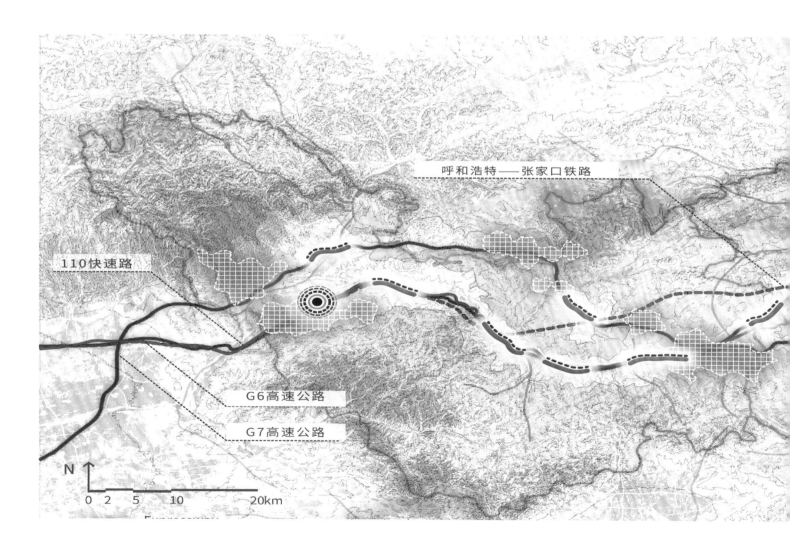

连接旅游资源的绿道

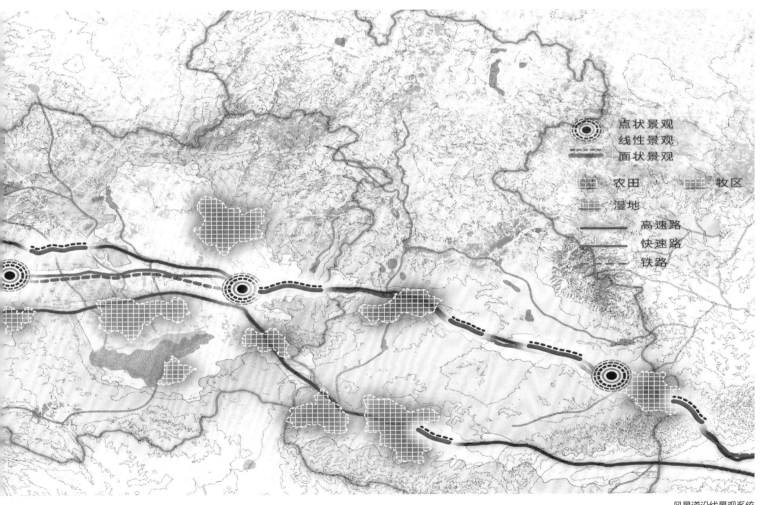

风景道沿线景观系统

1 公园城市与国土生态空间

活化利用文化遗产的4R模式
——山西晋城市陵川县文化风景道规划

4R Model of Activating and Utilizing Cultural Heritage
— Planning of Cultural Routes in Lingchuan County, Jincheng City, Shanxi Province

完成时间：2017年
建设地点：山西省晋城市陵川县
项目面积：1884.5 hm²
建设单位：山西省陵川县住房和城乡建设管理局

Time of completion: 2017
Construction site: Lingchuan County, Jincheng City, Shanxi Province
Project area: 1884.5 hm²
Construction unit: Administration of Housing and Urban-Rural Development of Lingchuan County, Shanxi Province

获奖信息：2021年国际风景园林师联合会亚非中东地区运动与休闲网络类卓越奖
Awards: Award International Federation of Landscape Architects Asia-Africa Middle East Region Award (IFLA AAPME) Sports and Network Excellence Award in 2021

主要设计人员：李雄　李方正　戈晓宇　等
Project team: LI Xiong, LI Fangzheng, GE Xiaoyu, et al

　　陵川县是一个文化底蕴深厚的城市，自古由太行山八陉之一的"白陉古道"带动发展。但由于古道历史功能的消失，陵川逐渐衰败，成为中国贫困城市之一，目前面临着环境、社会等复杂挑战。规划意在借鉴历史，基于古道提出"4R"目标，希望对文化遗产进行保护和利用，并综合生态修复、社区提升、产业策划多角度提出策略，实现以文化风景道为引擎的陵川老城复兴。项目构建了"文化源地识别—空间规划布局—线路营建实施"3个层级的规划框架。通过充分的调研，结合广泛的公众参与，梳理出陵川县内的31项文化遗产，并对其周边的自然组团、社区、废弃空间进行勘测，分析老城现状与线路要素的分布关系，在文化风景道的带动下，落地建设"一棵树"社区微绿地系统、"一公顷"城市公园系统与"一种生态技术"修复系统。规划线路结合修复历史古驿道，串联了陵川县大部分文化遗产点，形成17.6km的文化遗产展示线路和11.2km的古道更新体验线路，标志着陵川县文化遗产从零散到系统性保护利用的转变；在生态层面上，联系了北玄山、锦平郊野公园、卧龙岗、羊河等自然景观，修复了39.7hm²的绿地与城市废弃地；利用纵贯北路、梅园东街、开云街、棋山路等城市道路连接了14个社区和大部分城市开放空间，服务惠及约44800人，形成28.8km具有良好交通可达性与高服务覆盖性的城市游憩网络，为老城与古道的再次激活提供了可持续路径。

　　Lingchuan is a city with profound historical and cultural heritage. Since ancient times, "Bai Valley Pass", one of the eight passes of the Taihang Mountains, has driven the development of Lingchuan. However, as the historical function of the Pass fades away, it gradually falls into decay and becomes a city in poverty, facing complex environmental and social challenges. Considering the city's history background, the plan have proposed "4R" objectives to preserve and utilize the cultural heritages and put forward strategies from multiple angles including ecological restoration, community enhancement and industrial planning to realize the revival of the old city with the cultural routes as the engine. Through thorough research and extensive public participation, the project sorted out 31 cultural heritages in Lingchuan and surveyed their surrounding natural clusters, communities, and abandoned spaces to analyze the distribution relationship between the current state of Lingchuan and its cultural routes' elements. Driven by cultural route, "One Tree" community greenspace system, "One Hectare" urban park system and "One Eco-technology" restoration system are formed. The planned route combines the restoration of historical ancient post roads, connects most of the cultural heritage sites in Lingchuan, to form the 17.6 km long cultural heritage exhibition route and the 11.2 km long renewal experience route of the Pass. It marks the transformation of Lingchuan's cultural heritage from scattered to systematic protection and utilization. At the ecological level, the planed route has linked natural landscapes such as Beixuan Mountain, Jinping Country Park, Wolong Gang, Yang River, restored 39.7 hm² of green areas and urban waste land. The planned route connects 14 communities and most of the urban open spaces mainly by using urban roads such as North Zongguan Road, East Meiyuan Street, Kaiyun Street and Qishan Road etc. It serves approximately 44800 people and forms a 28.8 km urban recreation network with good accessibility and high service coverage, providing a sustainable path for the reactivation of the old county and the Pass.

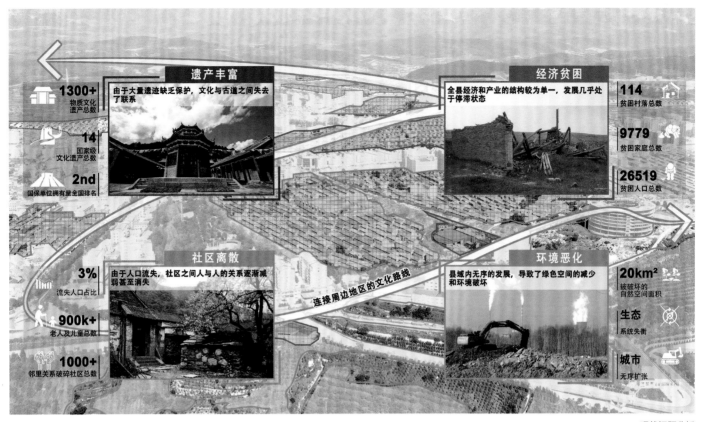

现状问题分析

陵川现状面临着许多方面的问题：大量遗迹缺乏保护，文化与古道失去联系；经济产业结构单一，发展几乎停滞；人口流失导致疏离的社区关系；无序发展导致了绿色空间减少和环境破坏。本规划对范围内的文化遗产价值、生态敏感性、开发适宜度、游憩潜力四个方面进行分析，明确遗产自身和周边发展潜力。

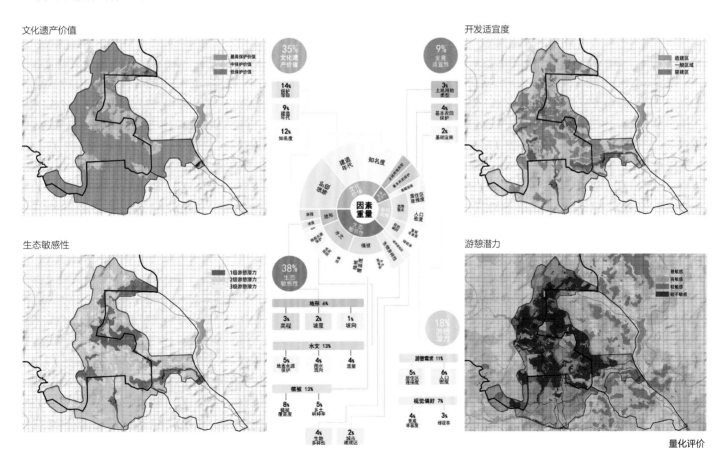

量化评价

文化遗产等级评价

从综合保护级别、基础设施建设、绿色空间建设三个方面对陵川县的文化遗产价值进行综合评估，并将其划分为4个等级，第一等级3个，第二等级2个，第三等级3个，第四等级1个。

对遗产周边的生态空间、社区进行勘测调查，结果显示8个文化遗产点周边环境恶劣，亟需修复；7个文化遗产点处于缺乏活力的社区之中，有待激活；5个文化遗产点分布在陵川县的偏远郊区，没有得到关注。

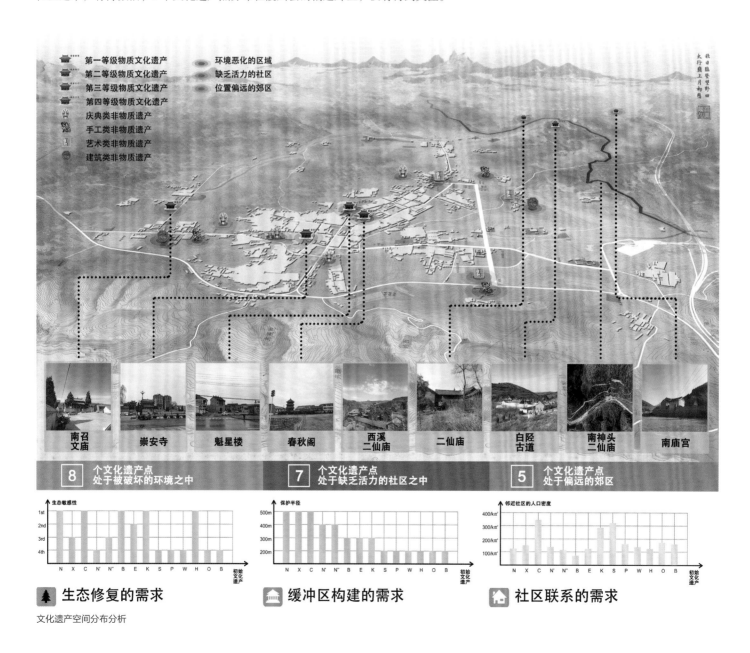

文化遗产空间分布分析

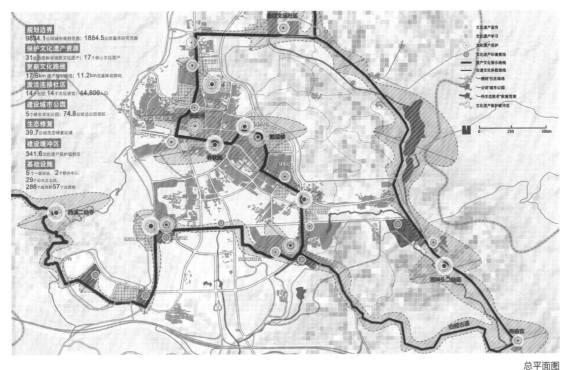

规划线路结合修复历史古驿道，串联了陵川县大部分文化遗产点，形成17.6km的文化遗产展示线路和11.2km的古道更新体验线路。

总平面图

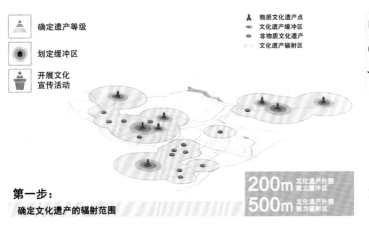

第一步：
确定文化遗产的辐射范围

文化焕活策略

根据遗产的保护等级，以物质文化遗产为中心划定缓冲区，以200m与500m为半径，明确其辐射范围。

第二步：
确定生态修复区域

绿色更新策略

在文化遗产缓冲区的范围内，通过生态修复技术优化现有生态空间，通过土地腾退创造更多的城市绿地。

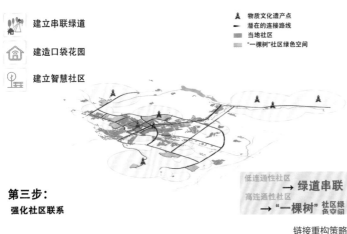

第三步：
强化社区联系

链接重构策略

依托现有线性的城市道路、河道、串联文化遗产点与临近社区，沿线以"一棵树"的方式增加社区绿色空间，激发活力。

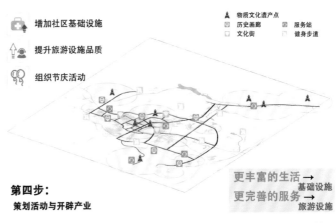

第四步：
策划活动与开辟产业

赋能运营策略

根据最终确定的文化风景道增加各类基础设施，加强活动策划，促进旅游业与休闲产业发展，丰富当地居民文化生活。

"一棵树"社区微绿地系统

文化风景道联系周边社区并融入居民生活,由社区文化微空间、社区游憩廊道、社区智慧网络,组成高服务覆盖性的"点、线、网"空间结构。围绕历史建筑建立"一公顷公园",划定了核心保护圈、建设控制区、景观统一区;并在规划的文化活动中心开展文化课堂等活动,传承非物质文化遗产。

"一公顷"城市公园系统

"一种生态技术"修复系统

修复古道沿线的生态环境，基于山地护坡与修复、水土保持与改良、湿地净化与健康三种营造手段，建立山地公园、径流缓冲公园与湿地公园。规划团队通过线上和线下两种途径，广泛采纳了陵川当地居民的建议，并从土地获得、空间营造、人员机制、合作基金四个方面为规划方案的实施提供保障。

多方参与机制

2

园林博览会与
事件性景观

Garden Exposition and
Event Landscape

城市事件（City Event）多指短期内发生的一系列重要活动的总和。包括奥运会、世界杯等庆典赛事，世园会、世博会等博览会展，以及战争疾病、自然灾害、重大事件事故等多种积极和消极的类型。事件性景观（Event Landscape）则是以大型城市事件的发生地为范围，使暂时性事件与恒久性环境建立紧密联系的一种景观类型。由于事件的独特性、筹办过程中集中人力物力投入、吸引大量游客和国际媒体关注等特点，对事件发生地产生短期或持续的集中影响，进而激活城市的更新与发展，成为推动城市基础设施和景观建设的重要途径。

园林园艺展作为典型的、积极的事件性景观，最早起源于欧洲，20世纪末开始在中国举办。经过20余年的发展，我国已成为举办园林博览会最为频繁的国家，也是每届博览会规模最大的国家，各类园林园艺博览会开展频率高、建设力度大、影响范围广，为举办城市带来了巨大的社会经济效益。

目前，国内举办的国家及国际级别的园林园艺博览会按照办会主体的不同主要划分为四种类型，分别是由国际园艺生产者协会（AIPH）和国际展览局（BIE）认定的世界园艺博览会、中国住房和城乡建设部主办的中国国际园林博览会、中国花卉协会主办的中国花卉博览会，以及全国绿化委员会主办的中国绿化博览会。自1987年第一届花博会以来，我国已相继举办32届国家级园林园艺博览会，涉及19省份25地市，总建设面积逾9000hm^2。

园林博览会等城市事件作为促进经济社会发展和城市环境更新的有利契机，具有加速城市景观和人居环境优化的积极带动效应。然而，当短暂的城市事件结束后，被遗留的景观却暴露出诸多问题。以园林博览会为例，出于对地区发展带动或对生态破坏修复示范等选址考虑，往往造成远城市选址、零基础建园等情况，对可达性和便捷性缺乏考虑，后续运营也不能得到充分保障，最终导致部分园区会后发展萧条，而这些问题在事件性景观中具有普遍的代表性。因此，风景园林需要充分认识事件性景观对城市发展的正负双重影响，从时间和空间维度出发合理谋划，从而带动整个城市的发展和景观建设。

时间维度：积蓄—发生—效应的全程动态融入

一般来说，大型事件都有从蓄积到发生再到产生效应的过程，作为城市建设的重要推动力，整个过程都与城市绿色空间构建有着密切关系。在会前积蓄期，有效借助城市发展的超前机遇，通过合理选址和科学规划，激活城市大型公共设施、交通网络、活力中心、开放空间、城市旧区等前瞻性建设和有机更新。在会事发生期，充分发挥城市事件的聚集效应，通过城市景观的魅力展示、城市形象的宣传推广、城市活动的热度升温，调动市民和游客的参与热情，激发城市整体发展活力。在会后效应期，持续利用事件性景观的广泛影响力，丰富完善市民的游憩和文化交流空间，实现城市吸引力和竞争力的持续提升。在事件发生全过程，特别是在蓄积期和效应期，需要充分发挥场地作为城市重要绿色空间和景观地标的主导角色，推动城市优势竞争力转化和区域协调发展。

空间维度：园区—周边—城市的绿色空间体系

作为大型城市绿地，诸如园林博览会等城市事件的举办场地是城市绿色空间体系中的重要斑块，发挥着充实城市功能、重塑景观格局、维护生物多样性和生态系统稳定性等关键作用。因此，园区景观应有机融入城市山水之间，保证绿色空间的连续性与自然景观的统一性，避免对自然生态的破坏和土地资源的浪费。在重视生态示范性的同时，考虑场地选址与城市功能区域的整体关系，兼顾交通可达性、场地吸引力，扩大可以为居民提供游憩体验和生态服务的功能空间。同时，需要将场地融入城市绿色基础设施、生态基础设施综合考虑，作为恒久性体系融入城市绿色空间，带动区域生态人居环境优化和经济社会发展。

在不断探索事件性景观紧密结合城市发展战略，构建城市绿色开敞空间系统，促进城市绿色活力激发的实践过程中，本章选择了4个实践项目，从不同角度解析事件性景观的塑造途径。

南阳洲际月季大会规划设计方案从城市生态文化体系角度出发，塑造城市生态文化展示窗口和景观地标，兼顾城市事件的临时性与集中效应，衔接过渡展会前后功能过渡。

河北省（秦皇岛）第二届园林博览会规划设计方案从城市生态基础设施角度出发，融入山海港城的生态景观风貌，塑造有生命的弹性景观系统，实现城市景观、功能、文化的融合，促进城市边缘地区的生态修复和可持续发展。

长沙市申办第十三届中国园林博览会的概念规划方案从城市双修及人居生态环境建设的角度出发，突出地域水景观价值，重塑城市生态肌理，通过事件性景观促进城市绿色空间质量提升，满足市民对于绿色低碳美好生活的福祉需求。

2012年第三届亚洲沙滩运动会主会场及公园规划设计方案从城市新型公共空间角度出发探讨事件性景观作为城市主义催化剂的方法路径，通过塑造兼顾赛前赛后使用需求的开放性、地域性、生态型的多功能景观，形成激发城市发展活力的绿色引擎。

City event refers to a series of important activities within a short period, including events like Olympic Games, World Cup, World Expo, etc. and wars, diseases, natural disasters, major events, and accidents. Event landscape is the landscape where there is a close connection between the temporary events and permanent environment within the range of place where big city events take place. Because of the uniqueness of the events, concentrated investment of manpower and materials during the preparation, a vast number of attracted tourists and highlight of international media, there is a short-term or constantly concentrated influence to activate the updating and development of cities, which is an important approach to push the infrastructure and landscape construction in the city.

As a typical and positive event landscape, the garden expo was originated in Europe and had been held in China since the end of the 20th century. With more than 20 years of development, China has become the one that holds garden expos the most frequently with the biggest scale for each expo. With high holding rate, tremendous investment and wide influences, all kinds of garden expos have brought tremendous social and economic benefits to the host cities.

By far, the national and international garden expos held by China fall into four categories according to the different hosting subjects, including the World Horticultural Exposition certified by the International Association of Horticultural Producers (AIPH) and International Exhibitions Bureau (BIE), China International Garden Flower Expo held by the Ministry of Housing and Urban-Rural Development of the People's Republic of China, Chinese Flora Expo held by China Flower Association and China Aflorestation Expo held by National Aflorestation Environmental Protection Commission. Since the first China Flower Expo in 1987, China has successively held 32 sessions of national garden expos, involving 25 cities and 19 provinces and a total area of over 9000 hectares.

Garden expos and other city events have been a new advantage to promote economic and social development and update the urban environment, which provide positive drive to accelerate the urban landscape and improve the living environment. However, when the momentary city events are finished, there are so many problems in the landscapes left behind. Take garden expos as an example, the remote location of city and construction with zero foundation would be caused for the consideration of site selection for regional development or ecological damage restoration demonstration, so there is a lack of consideration on the accessibility and convenience. The follow-up business is also not given enough attention so that some parks come out to be in a sluggish development, and these problems are widely represented in the event landscapes. Hence, the positive and negative impact of the event landscape on urban development should be taken into full consideration, so as to have a reasonable plan from the temporal and spatial dimensions to boost the development and landscape construction in the whole city.

Temporal Dimension: A whole-process dynamic integration of saving, happening and affecting

Normally speaking, the big events would have a process from saving, happening and finally to affecting. As an important promoter of urban construction, the whole process has a close relationship with the construction of green space in the city. In the saving period before the events, it could effectively make use of the advancing opportunity of urban development to activate the forward-looking construction and innovative renewals such as urban large-scale public facilities, transportation network, vitality centre, open space, old urban area through reasonable address selection and scientific planning. During the happening of an event, it should have a full play of the aggregation effect of city events. Through the charming presentation of urban landscape, publicity of urban image, warming up for urban activities, it could activate citizens and tourists' passion to participate in the activities and trigger the overall development energy of the city. In the effecting period after the events, the wide influence of event scenery should be constantly used to enrich and improve citizens' recreation and cultural exchange space so as to achieve the continuous growth of the attraction and competitiveness of the city. During the whole process of the events, especially in the saving and effecting periods, the venue should play a vital role for the important green space and landscape landmark of the city, and it would catalyze the competitive advantages and promote regional coordinated development in the city.

Spatial Dimension: A green space system of park-surrounding-city

The hosting venues of such city events as garden expos are important blocks of the green space system of the city as a large-scale urban green land, playing a key role in enriching urban functions, reshaping landscape pattern, maintaining biodiversity and ecological stability. Hence, the garden landscape should be systematically integrated into the natural environment of the city to ensure that the continuity of green space and unity of natural landscape and avoid the destruction of natural ecology and the waste of land resource. With significance attached to ecological demonstration, the overall relationship between venue selection and urban functional areas should be taken into consideration with regards to the accessibility and venue attraction, so as to expand the functional space to provide recreation experience and ecological services for residents. At the same time, the venue should be integrated into urban green and biological infrastructures and hence as a permanent system in the urban green space drive the improvement of living environment and economic and social development in the region.

While intensively elaborating on the relationship between event landscape and urban development strategy, the practicing process of building up an urban green open space system to advance the green vitality of the city, this chapter chooses 4 cases to explain the shaping approaches of event landscape from different aspects.

Planning and design for the WFRS Regional Convention begins from the aspect of urban ecological culture system. It shapes the display window and landscape landmark of the urban ecological culture, combines the temporality and concentration effect of urban events, and connects the functional transition before and after the exhibition.

Planning and design for the Second Garden Expo of the Hebei Province (Qinhuangdao) begins from the aspect of urban biological infrastructure. It is integrated with the ecological landscape of the coastal city and shaped with a flexible landscape system with life to achieve the integration of urban landscape, function, and culture, so as to advance the ecological restoration and sustainable development in the urban fringe.

Conceptual planning for the Changsha City to apply for holding the 13th China Garden Expo begins from the aspect of the ecological and urban restoration of the city and the construction of the ecological and living environment. It highlights the regional water landscape value and reshapes the urban ecological texture. The green space quality of the city would be boosted through event landscape to satisfy citizens' welfare needs for green, low carbon and better life.

Planning and design for the main venue and park of the 3rd Asian Beach Games in 2012 begins from the aspect of new urban public space to discuss the methods and ways to make event landscape as a catalyst of urbanism. It is formed as a green engine to activate urban development vitality by reshaping and caring for the open, regional, and ecological multi-functional landscape before and after the competition.

为城市绽放的花园
——2019年南阳世界月季洲际大会博览园

The Garden Blooming for the City
— 2019 WFRS Rose Regional Convention EXPO Park in Nanyang

建成时间：2019年	Time of completion: 2019
建设地点：河南省南阳市城乡一体化示范区	Construction site: Demonstration Zone for Urban-Rural Integration, Nanyang City, Henan Province
项目面积：83.1 hm²	Project area: 83.1 hm²
建设单位：南阳市城乡一体化示范区管委会	Construction unit: Administrative Committee of Nanyang

获奖信息：2021年国际风景园林师联合会亚非中东地区公园与开放空间类卓越奖
世界月季联合会世界月季名园奖

Awards: Award International Federation of Landscape Architects Asia-Africa Middle East Region Award (IFLA AAPME) Parks and Open Space Excellence Award in 2021
WFRS World's Famous Rose Garden

主要设计人员：李雄 张云路 林辰松 孙漪南 胡楠 赵鸣 段威 等
Project team: LI Xiong, ZHANG Yunlu, LIN Chensong, SUN Yinan, HU Nan, ZHAO Ming, DUAN Wei, et al

本项目是2019年第十八届世界月季洲际大会的举办场所。世界月季联合会（WFRS）于1968年在英国伦敦成立，其宗旨是举办国际月季会议，并作为月季研究的信息交流中心。世界月季联合会每三年举行一次大型的国际会议，将世界各地的月季爱好者和专家聚集在一起，进行花园参观和专家讲座。

南阳月季栽植历史悠久，月季文化浓郁，月季产业突出，是中国的月季之城。本项目以中国传统山水园为基础，通过月季单体品种展示、月季群体效果展示以及月季元素景观表达构建了全园月季主题系统，营造了具有浓郁的月季主题氛围以及中国特色的月季博览园，圆满完成了月季文化展示、月季研究促进、月季产业交流以及市民休闲游憩的办会任务。并被世界月季联合会评为"世界月季名园"，极大彰显了南阳城市形象与月季文化。

本项目为城市增加了一处月季主题绿色空间，优化了城市绿色空间体系，形成了城市月季主题绿色空间系统；促进了月季产业规模扩张与品质提升，带动了城市经济发展；满足了市民日常休闲游憩的需求，使月季文化更加深入人心。

The project is the exhibition park of the 18th WFRS regional conference that held in Nanyang, China, in 2019. The World Federation of Rose Societies (WFRS) was founded in 1968 in London, UK. Its stated purpose was to hold international rose conferences and act as a clearing house for rose research. Every three years the WFRS holds major international conventions, bringing together rose enthusiasts and experts from around the world with garden visits and expert lectures.

Nanyang is the rose city of China, with a long history of rose planting, abundant rose culture and outstanding rose industry. The theme of this park focuses on rose display: display of individual rose varieties and display of rose group-landscape, and expression of rose elements. The project creates a city park with strong rose theme atmosphere and Chinese characteristics based on Chinese traditional landscape garden, which successfully completes the task to display rose culture, promote rose study development, promote rose industry exchange and provide public leisure space. Moreover, the project is conferred the title of "World's Famous Rose Garden" by the World Federation of Rose Societies, which greatly highlights the city image and rose culture of Nanyang.

The project offered a rose themed green space for the city, which optimizes the urban green space system, forms the urban rose themed green space system, promotes the expansion and quality of rose industry, drives the urban economic development, meets the daily leisure needs of citizens, and makes rose culture rooted in the hearts of the people.

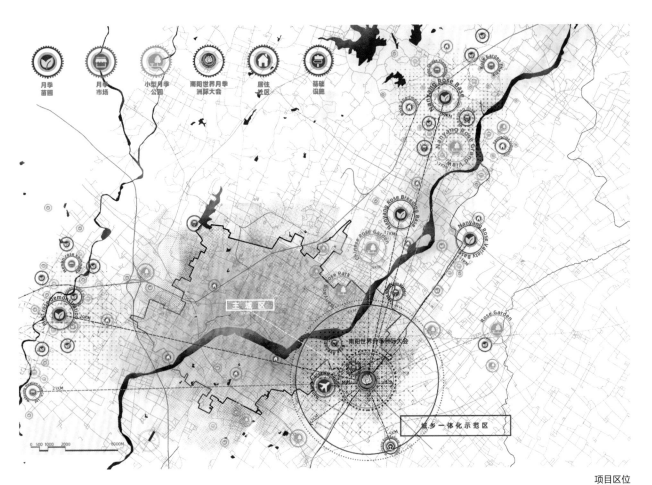

项目区位

图例
1. 南阳东道主展园
2. 卧龙书院
3. 5A级游客服务中心
4. 世界月季大舞台
5. 城市展园
6. 品系月季园
7. 岩石园
8. 蔷薇属植物专类园
9. 色彩月季园
10. 精品月季园–芳香园
11. 中国自育品种区
12. 东湖
13. 西湖
14. 花堤
15. 儿童园
16. 盆景园
17. 花街
18. 授赏月季园
19. 树状月季园
20. 主停车场
21. 次停车场
22. 主入口
23. 次入口
24. 花山
25. 花廊
26. 花海
27. 飞机跑道
28. 花洲花岛

总平面

为彰显南阳的景观特色，将中国传统的造园技艺融合到月季展示空间的塑造中，通过山形、水系组织塑造不同展示区域、展示空间，形成以中国传统自然山水园为基底的月季展示体系空间骨架。功能分区共分三大区域，分别为东园——月季大会主展区（47.8hm^2）、西园——月季花圃大地艺术区以及月季花海游赏区（35.3hm^2）。

2 园林博览会与事件性景观 65

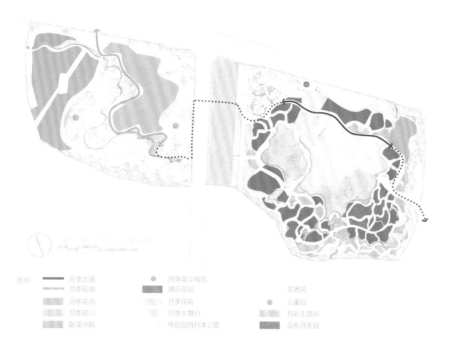

月季群体效果展示

月季花海

月季花山

月季花廊

月季元素景观表达

月季花伞构筑

月季花街

月季单体品种展示

专类花园 >>

🌹 月季大道

🌿 品系月季园

⭐ 卧龙书院

展示花园 >>

🌱 中国自育月季之路

🌸 月季大舞台
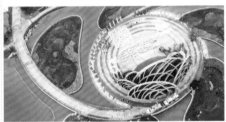

🏛 城市展园

主题花园 >>

🌈 色彩主题园

🐝 芳香园

🎈 儿童园

月季展示体系

基于中国传统山水空间的月季展示体系

2 园林博览会与事件性景观

月季花廊　　　　　　　　　　　　　　　　　月季花廊内部

　　本项目月季群体效果展示主要位于西区，以大地艺术的方式打造月季花山、月季花海。以大尺度空间为载体，侧重于月季宏观场景的营造，以突出月季色彩美感。

　　受限于建设高度控制，西区难以设计较高的地形，且无法种植高大乔木。因此，西区以缓坡地形为主，使用月季作为主体植物材料，结合宿根花卉，通过大地艺术的手法，不仅打造了壮阔的月季景观，还创造了空中观景的可能，将原本因邻近机场而产生的建设劣势变为景观特点。

月季花海

　　建筑、山石、植物、景墙、景观小品等，都作为设计元素融入项目多样性景观当中，形成了功能多样、风格各异的休闲空间，极大地丰富了南阳市民的日常休闲娱乐生活。

大地艺术区

观景台

通过本项目的建设,南阳月季产业优势得到了充分发挥,扩大了月季产业规模,推动了传统月季产业向创新、高质量的精品式产业发展并带动了周边的乡村振兴。

本项目被世界月季联合会评为"世界月季名园"。通过建设提升了南阳在全球的城市形象与知名度,打开了南阳月季品牌的国际市场,丰富了南阳市民的日常休闲途径,强化了月季文化在市民心中的认同感。

观景台远景

棕地生态修复的城市绿色引擎
——河北省第二届园林博览会（秦皇岛）规划设计

Urban Green Engine for Brownfield Ecological Restoration
—— Planning and Design of the Second Garden Expo Park (Qin Huangdao) of Hebei Province

建成时间：2018 年	Time of completion: 2018
建设地点：河北省秦皇岛市	Construction site: Qinhuangdao City, Hebei Province
项目面积：137 hm²	Project area: 137 hm²
建设地点：秦皇岛市园林局	Construction unit: Qinhuangdao Municipal Bureau of Landscape Architecture

获奖信息：2019 年国际风景园林师联合会亚非中东地区公园及开放空间类荣誉奖、文化及城市景观类荣誉奖
2020 年国际风景园林师联合会亚非中东地区奖雨洪管理类荣誉奖、经济活力类荣誉奖
2021 年中国风景园林学会科学技术奖规划设计一等奖
2021 年北京风景园林学会规划设计一等奖

Awards: Award International Federation of Landscape Architects Asia-Africa Middle East Region
Award (IFLA AAPME) Parks and Open Space Honorable Mention in 2019; Cultural and Urban Landscape Honorable Mention in 2019
Award International Federation of Landscape Architects Asia-Africa Middle East Region
Award (IFLA AAPME) Flood and Water Management Honorable Mention in 2020; Economic Vitality Honorable Mention in 2020
First prize of Planning and Design of science and Technology Award of Chinese Society of Landscape Architecture in 2021
First prize of Planning and Design of Beijing Society of Landscape Architecture in 2021

主要设计人员：李 雄 姚 朋 戈晓宇 肖 遥 董 丽 郝培尧 林 洋 段 威 葛韵宇 等
Project team: LI Xiong, YAO Peng, GE Xiaoyu, XIAO Yao, DONG Li, HAO Peiyao, LIN Yang, DUAN Wei, GE Yunyu, et al

河北省第二届园林博览会在中国河北省秦皇岛市举办，位于城西经济技术开发区栖云山片区东部，占地约 137hm²，贯穿宁海大道和天津至秦皇岛铁路客运专线。场地位于城市边缘，距离城市中心 11km，发展滞后。所在的栖云山片区，过去频繁的采矿活动对环境造成了严重的破坏，造成了山体、植被、农田、水环境等一系列的环境问题。

该项目旨在通过城市活动和建设园林博览会来解决栖云山片区的环境污染问题和绿色开放空间的缺乏状况。以事件景观为绿色引擎，推动周边地区建设，实现城乡一体化发展。在专业层面，项目通过建立一个园林盛会的平台，展示园林艺术和技术，促进行业交流；在生态层面，项目通过切实可行的措施，实现了栖云山片区矿坑和山沟的生态修复；在公共层面，通过规划会议期间和会议之后的活动，为持续的互动和分享提供了一个绿色的开放空间。

The Second Garden Expo Park of Hebei Province is located in Qinhuangdao City, Hebei Province, China. It is located in the eastern part of Cloud Mountain District, the economic and Technological Development Zone in the western part of the city. It covers 137 hm² through Ninghai Avenue and Tianjin-Qinhuangdao Railway Passenger Line to the east of the site. Located at the edge of the city, the site is 11 km away from the central city, and its development is lagging behind. Cloud Mountain area, where the Garden Exposition Park of Hebei Province is located, was seriously damaged by frequent mining activities in the past, which caused a series of enviroment problems, such as mountains, vegetation, farmland and water environment.

This project aims to solve the environmental pollution problem and the lack of green open space in Cloud Mountain area by means of urban events and the construction of garden exposition park. Taking the event landscape as a green engine to promote the construction of surrounding areas and realize the integration of urban and rural development. At the professional level, the project promotes industry exchanges by creating a platform for landscape grand events, displaying landscape art and technology; at the ecological level, it realizes the ecological restoration of mining pits and mountain gullies in Cloud Mountain area through practical measures; and at the public level, it provides a green open space for continuous interaction and sharing through the planning of activities during and after the meeting.

秦皇岛位于中国渤海西部，是中国最先开放的沿海城市之一，但城市西北部发展严重滞后，同时海陆发展极不平衡。

秦皇岛第二届园林博览会位于秦皇岛市西部栖云山片区，位于城市边缘，希望借助城市事件和园林博览会的建设，促进秦皇岛西部的发展，实现秦皇岛东西部的均衡发展。

区位图

总平面图

2 园林博览会与事件性景观

矿山生态修复

建成实景

72　理地营境　生态文明建设背景下风景园林实践

秋景园

建成实景

2　园林博览会与事件性景观

建成实景

实景鸟瞰

2 园林博览会与事件性景观

"水韵园博、圆梦人居"的城市双修典范
——第十三届中国园林博览会长沙市申办概念规划

A Model of City Betterment and Ecological Restoration Programs with "Rhyming Water Garden Expo, Living Dream Human Habitat"
— The 13th China International Garden Expo Conceptual Planning for Changsha's bid

完成时间：2017 年	Time of completion: 2017
建设地点：湖南省长沙市雨花区	Construction site: Yuhua District, Changsha City, Hunan Province
项目面积：260 hm²	Project area: 260 hm²
建设单位：长沙市园林管理局	Construction unit: Changsha Landscape Bureau

主要设计人员：李 雄 冯 潇 张云路 肖 遥 尹 豪 李 翅 殷伟达 于长明 等
Project team: LI Xiong, FENG Xiao, ZHANG Yunlu, XIAO Yao, YIN Hao, LI Chi, YIN Weida, YU Changming, et al

　　长沙市申办园林博览会的概念规划以我国全面推进城市双修，全面建成小康社会及长沙市创建生态园林城市为战略背景。拟定园博园主会场位于长沙市雨花区潭阳洲，规划面积260hm²，主会场外围设置城市发展协调区1890hm²，并设有柏加镇分会场，主会场与分会场之间的浏阳河段将作为一条特色的水上通道连接两个会场。规划以"水韵园博，圆梦人居"为办会主题，提出"全新、全民、全景、全域、全时"的办园理念，力求通过园博会来促进城市园林绿化质量的提升和园艺的发展，使园林真正融入生活，倡导绿色低碳生活，让大众享受绿色福利、提升幸福感的同时，推进生态文明建设，构建美丽人居环境，引导城市绿色健康发展。

The Conceptual Planning for Changsha's Bid for the 13th China International Garden Expo is based on the strategic background of China's comprehensive promotion of urban double repair, the comprehensive construction of a well-off society and the creation of an ecological garden city in Changsha. The main venue of the garden expo is proposed to be located in Tanyangzhou, Yuhua District, Changsha, with a planned area of 260 hm². An urban development coordination area of 1890 hm² is set up on the periphery of the main venue, and a branch venue is set up in the town of Bojia. The section of the Liuyang River between the main venue and the branch venue will serve as a distinctive waterway linking the two venues. With the theme of "Rhyming Water Garden Expo, Living Dream Human Habitat" and the concept of "New, National, Panoramic, Territorial, All Time", the planning aims to promote the improvement of the quality of urban landscaping and the development of horticulture through the Expo, so that gardens can truly integrate into life. The Expo advocates green and low-carbon living, lets the public enjoy green benefits and enhance their sense of well-being, while promoting ecological civilisation, building a beautiful living environment and guiding the green and healthy development of the city.

主会场用地概况

潭阳洲位于长沙市中心区东部，东靠临湘山，浏阳河三面环绕，与岳麓山、橘子洲遥相呼应，展现了长沙"山水洲城"的特色。潭阳洲交通条件便利，距离长沙火车站10km、高铁站7km、黄花机场12km。用地路网发达，连接高铁长沙南站和黄花国际机场的中低速磁悬浮快线和机场高速从潭阳洲穿过。用地地势平坦，沿浏阳河设有堤防，河堤长约6km，堤顶宽度为10m，堤顶较场地平均高4~6m，规划用地总面积260hm^2。

潭阳洲发展史

潭阳洲古称梨江湾，是浏阳河流域第一大湾。这里曾是"梨江八景"中的"梨江双渡"。经过人们数千年的劳作，洲上逐渐形成了"圩垸"的村落肌理。后来随着居住人口的增长，人们在潭阳洲上大量建造住房，逐渐形成长沙市最大棚户区。2015年，长沙市正式启动了该棚户区的拆迁工作，截至2015年底，长沙市已拆除潭阳洲违章建筑百万平方米，洲岛违章建筑得到全面清除，也为长沙市申办园林博览会打下了基础。

1. 西主入口　2. 南主入口　3. 南次入口　4. 东入口（紧急）　5. 北主入口　6. 市民生活主题展园　7. 国际大师园
8. 东道主展园　9. 九大流域城市展园　10. 国际展园　11. 棚改绿示范园　12. 企业园　13. 国内大师园　14. 园衍园
15. 渔村夕照　16. 江天暮雪　17. 洞庭秋月　18. 山市晴岚　19. 平沙落雁　20. 烟寺晚钟
21. 远浦归帆　22. 潇湘夜雨　23. 九嶷山　24. 云梦湖　25. 主展馆（地铁接驳）　26. 温室

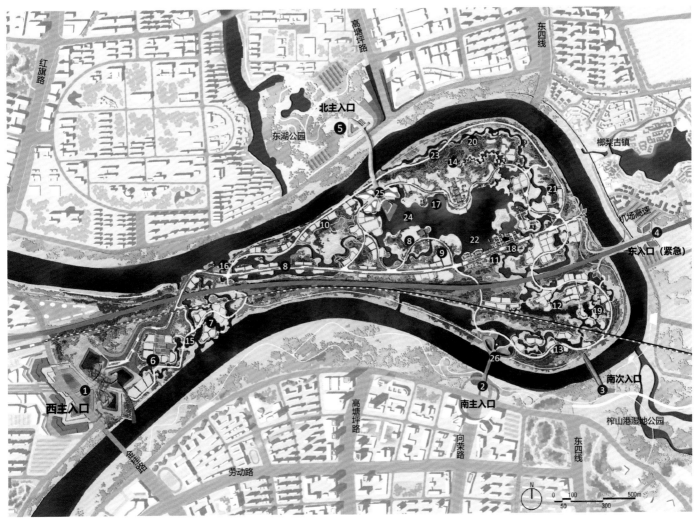

设计总平面

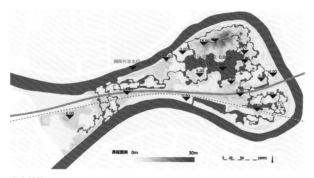

山水结构

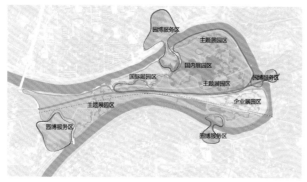

功能分区

水体生态净化分区

办展特色

规划以"水文章、园梦想、湘意境"为特色。重点突出：

水文化——充分利用浏阳河的资源与潭阳洲的地貌，强调地域水景观的特征，突出水生态的价值，践行水生态文明建设。

园梦想——实现城市修补与生态修复，满足人民群众日益增长的绿色生活需求，体现城市绿色基础设施的价值。

湘意境——建设充满湘风、湘景、湘土、湘情的园博园，重视本土文化传承，彰显地域景观特色，让湖湘文化成为展会的本底特色。

山水结构

方案以湖南山水特色的典型代表——"九嶷山""洞庭湖"为概念营造园区空间格局，形成了"九嶷宗脉、洞庭汇芳"的山水骨架。在此基础上，通过在洲内引入浏阳河水，形成一条完整的内部水脉体系，承载了水生态、水游憩、水文化的综合功能，形成园区山水景观和展园体系的重要线索。

展园结构

依托山水结构，规划方案形成了"两轴两环、一脉五区"的总体结构。"两轴"是指贯通整体园区的"人居绿色梦想轴、园博山水景观轴"。"两环"是指沿水系布局的"万众同心创绿环、万里江山锦绣环"。"一脉"是指串联园区的潇湘水脉。"五区"是指沿潇湘水脉分布的国内展园区、国际展园区、主题展园区、企业展园区、园博服务区。

主会场鸟瞰

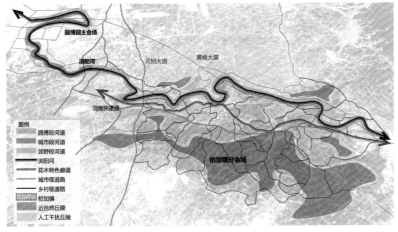

规划结构

协调区总体规划

为实现"一园带一区,一区范一城"的城市发展目标,规划在主会场外围设置面积 1890hm² 的城市发展协调区,将园博会与周边城市绿色生态网络、基础设施体系、城市产业与城市生活紧密结合起来。

柏加镇分会场与浏阳河绿色风光带

规划突破主会场空间限制,充分利用长沙市柏加镇绿色产业基地的有利条件,将柏加镇设立为园博会分会场,打造集观光体验、绿色人居、科普展示、科研示范、生产销售为一体的园博主题特色小镇。

并且利用浏阳河串联两个会场,形成了 50km 长的以"山-水-林-田"自然资源为支撑的绿色风光带。

协调区总体规划

2 园林博览会与事件性景观

"事件景观"新公共空间作为城市主义的催化剂
——2012年第三届亚洲沙滩运动会主会场及公园规划设计

New Public Space of "Event Landscape" as the Catalyst of Urbanism
— The 3rd Asian Beach Games Main Venue and Park

建成时间：2012年	Time of completion: 2012
建设地点：山东省烟台市海阳新区	Construction site: Haiyang New District, Yantai, Shandong Province
项目面积：405 hm²	Project area: 405 hm²
建设单位：海阳市园林局	Construction unit: Haiyang Municipal Bureau of Park
获奖信息：2018年国际风景园林师联合会亚非中东地区经济活力类卓越奖	Awards: Award International Federation of Landscape Architects Asia-Africa Middle East Region Award (IFLA AAPME) Economic Vitality Excellence Award in 2018
2018年中国风景园林学会优秀规划设计项目奖二等奖	Second prize of Excellent Planning and Design Project of Chinese Society of Landscape Architecture in 2018
主要设计人员：李雄 郑曦 李运远 冯潇 姚朋 戈晓宇 张云路 等	Project team: LI Xiong, ZHENG Xi, LI Yunyuan, FENG Xiao, Yao Peng, GE Xiaoyu, ZHANG Yunlu, et al

第三届亚洲沙滩运动会在中国海阳市举办，其主会场和奥林匹克公园的设计是由风景园林师主导的多专业合作设计模式，景观设计长期目标是以事件景观作为触媒，推动赛后的城市开发，并作为未来城市新区的公共空间系统，为赛会后场地持久的活力和可持续发展作出贡献。

亚洲沙滩运动会主会场及公园设在亚沙城新区中，总面积约405hm²。主会场和奥林匹克公园以林地、广场、公园、湖面、大海以及主会场绿岛剧场为结构，构建了一个面向城市开放的景观系统。该公园的建立以亚沙会这一洲际赛事为契机，把大型体育事件与城市景观相互融合，旨在创建一个开放的、活跃的公共空间，作为区域发展的促进力，推动滨海小城市大发展的梦想。

The 3rd Asian Beach Games, is one of Asia's four state-level sporting events, held in Haiyang City, Shandong Province, China, on June 16-23, 2012. The design of its main venue and the Olympic Park is a multi-disciplinary cooperation design model led by landscape architects. The long-term goal is an event landscape as a catalyst to promote the city development after the Games, and as a future urban landscape system to contribute for the vitality and sustainable development.

The Asian Beach Games Main Venue and Park are situated in the new district of the Yasha city, with a total area of about 405hm². The Main Venue and the Olympic Park build a landscape-oriented urban open space system, including woodland, squares, gardens, lake, as well as the eco-island in Yellow Sea. The establishment of the park takes the ASA as an opportunity to integrate large-scale sports events and urban landscape with the aim of creating an open and active public space as a driving force for regional development and promoting the dream of large-scale development of small coastal city.

总平面

1区：亚沙会生态岛主会场区
1. 主会场广场（紧急疏散，安保）
2. 观众入口广场
3. 临时停车场（赛会期间，媒体，装备）
4. 绿色剧场（主会场）
5. 入口（观众，运动员与媒体，VIP）
6. 火炬塔
7. 泻湖
8. 观景平台
9. 黄海
10. 防浪堤

2区：奥林匹克公园区
11. "海岛"主题园
12. 庆典草坪
13. "阡陌"主题园
14. 溪流（收集地表径流用于浇灌）

3区：中心广场区
15. 会议中心
16. 中心广场
17. 乡土树木区
18. 亚沙环岛

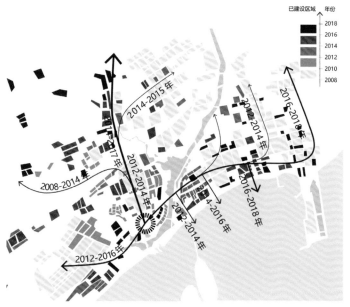
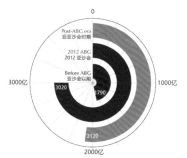

房地产发展

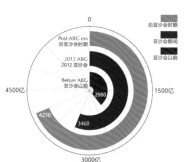

旅游业发展

城市建设时间轴

规划设计致力于以亚沙会这一洲际赛事为契机，把大型体育事件与城市景观相融合，作为未来城市新区的公共空间系统，留下亚沙会印记，推动滨海小城市大发展的梦想。

由于亚洲沙滩运动会，海阳新区的城市结构得到了改善。曾经作为绿轴的地块，在接下来的6年里引领了新区的成长。

亚洲沙滩运动会公园位于新区中轴线上，带动周边地区的成长和发展，特别是沿海旅游产业的完善，形成了城市事件驱动的"亚洲沙滩运动会效应"。

近年来，在"亚洲沙滩运动会效应"的作用下，城市的地产和旅游经济稳步发展，形成了"后亚洲沙滩运动会"时期模式。

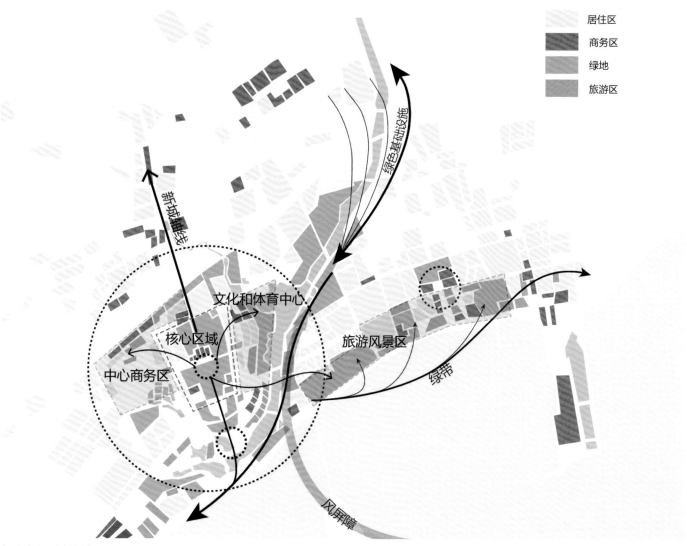

"亚洲沙滩运动会效应"

生态岛绿色剧场作为主会场的方案取代了原有城市总体规划中在亚沙城东部区域建设标准体育场的方案，节省了约70%的建设费用。

策略1——体育场重建

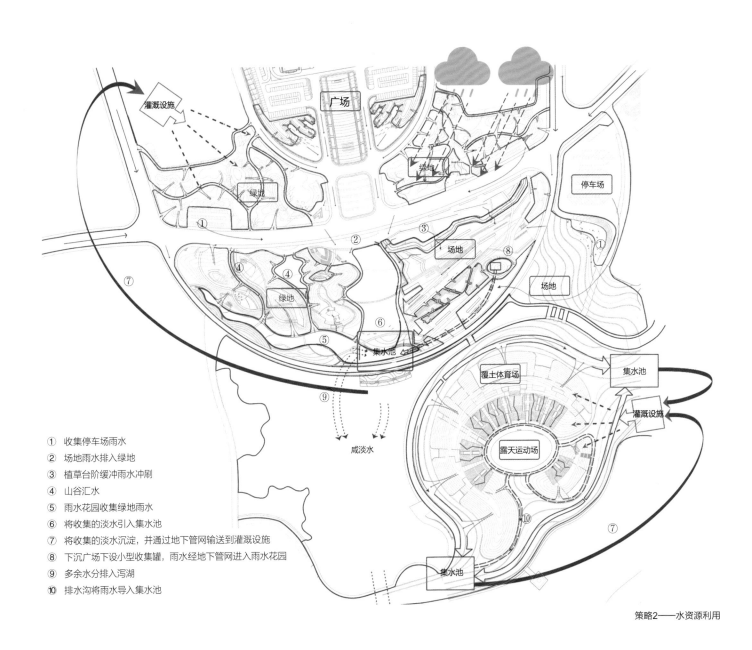

① 收集停车场雨水
② 场地雨水排入绿地
③ 植草台阶缓冲雨水冲刷
④ 山谷汇水
⑤ 雨水花园收集绿地雨水
⑥ 将收集的淡水引入集水池
⑦ 将收集的淡水沉淀，并通过地下管网输送到灌溉设施
⑧ 下沉广场下设小型收集罐，雨水经地下管网进入雨水花园
⑨ 多余水分排入泻湖
⑩ 排水沟将雨水导入集水池

策略2——水资源利用

2 园林博览会与事件性景观

地形与主赛场的地形相呼应，增加了景深

结合微地形设置多层种植池，设置景观景墙

主会场是公园中最大的开放空间

观众入口广场景墙，其灵感来自波浪的韵律

84　理地营境　生态文明建设背景下风景园林实践

从"船"茶室到奥林匹克公园的"岛"花园,地形、山谷和桥梁形成丰富的空间层次和体验

奥林匹克公园的小溪、绿色岛屿和桥梁。岛屿之间形成的山谷汇集了流入溪流的地表径流,连接岛屿之间的桥梁是海洋的隐喻

3

植物园
Botanical Garden

植物是园林中最重要的活体要素，也是人们生活中不可或缺的自然资源。据统计，世界上共有30多万种高等植物，我国仅种子植物就超过25000种，其中乔灌木种类约8000多种。人们对植物世界探索的脚步从未停止，这种探索逐渐演变为一种收集行为，从事这一活动的人也常被形象地称为"植物猎人"。可以说植物收集的过程充满了许多未知和不确定性，对于世界植物学的发展和植物园的建设也起到了重要的作用。

植物收集历史与植物园的发展一脉相承，国内外很多古老的苑囿园林中都收集并栽培了大量的专类植物，如秦汉时期的皇家园林上林苑、5世纪欧洲的修道院等，当时收集的植物多为药用植物以及具有一定经济价值的植物，用作观赏的植物还只占很小一部分。

16世纪，意大利的比萨植物园和帕多瓦植物园建成，植物园的功能由单纯的收集和栽培经济作物逐渐向植物引种、科研教学转变，此时植物园的结构始终延续着中世纪修道院花园的规则式格局。

进入17世纪，随着植物分类学的发展以及文艺复兴对文化与经济的冲击，一场世界植物掠夺大战缓缓拉开，植物园内也专门开辟了用于研究收集来自世界各地植物资源的区域，植物园的结构也更加灵活自然。然而面对来自不同地区的植物，如何成功引种并栽植成活成为一个新课题被提出，也由此促进了设施栽培技术的发展。工业革命为这一发展提供了可能，一些大型的热带温室建成了。自此，从世界各地竞相引种，收集珍奇的植物成为植物园的首要任务。

19世纪开始，城市建设迅猛发展，公共绿地对生长性状良好的植物需求增加，植物园也逐渐承担起了引种驯化、选育及试种植物新品种的任务，有些植物园还专门开辟了引种驯化区，使植物园向支撑产业发展又迈进了一步。我国也是从这一阶段开始出现了以科研为主的植物园，完成了古典园林中植物专类种植区向公共科研性质植物园的转型，然而，与先进水平的植物园相比，仍存在较大的差距。

进入20世纪后，人类逐渐意识到快速的发展和对自然界的过度索取正在使大量的植物种类破坏加剧，据推测到21世纪末，生态环境的恶化将导致1/2的植物面临生存威胁，超过2/3的维管植物可能完全消失，人们逐渐意识到植物园对珍稀濒危植物保护研究具有重要意义，将植物园视为挽救植物物种免遭灭绝的"方舟"。

今天的植物园在进行自身建设的同时，还设立了种质资源共享机制，搭建了更为综合的科研交流平台。至今加入国际植物园保护联盟的植物园数量已有3600多所，在保证一般公园绿地功能的基础上，还具有对活植物进行收集和记录管理，使之用于保护、展示、科研、科普、推广利用的独特属性。

每座城市都需要一个属于这座城市的植物园，《国家生态园林城市标准》规定"地级市至少有一个面积40hm^{2}以上的植物园"，促使中国植物园事业蓬勃发展起来。2020年《植物园设计标准》发布，对于保证植物园设计质量，全面发挥植物园的综合效益有着重要意义。

生态文明建设的提出，使中国社会的价值观有了全新的转变，顺应自然、保护自然、利用自然成为新的发展趋势。植物园的建设与发展，不仅是为了实现建设数量上的飞跃，更要在地理分布的均衡性、植物收集的多样性、功能类型的完整性、科学研究的独特性、交流合作的持续性等方面发挥更大的作用。随着生态文明体制改革，对美丽中国的建设愈加重视，植物园将成为未来加强生态文明建设的生动实践。

新时代科技的进步与人工智能的发展，也为植物园的建设与发展创造了新契机。本章选择了4个实践项目，立足于当今世界特定的环境条件，提出了植物园的建设策略，将对捍卫宝贵的绿色资源、延续人类文明产生重要的意义。

国家植物博物馆园区总体规划探索了"馆园一体，景研结合"的创新型植物展陈模式，打造了由山水、动植物、建筑和景观相互映衬的，活体生物与人类和谐共存的综合性展示空间。

中原黄河植物园规划设计作为首个系统收集培育展示黄河流域植被的植物园，对我国中原地区的南北过渡带地区植物进行了系统分类展示，提出了境域式植物景观展示策略，并赋予了植物园多重参与功能。

石家庄植物园在近自然和生物多样性等设计理念的引导下，专注于展示河北地区的原生生物群落，作为生物多样性保护基地、华北地区重要的种质资源库、科普教育与生态旅游相结合的教育基地，充分发挥了综合性植物园的示范作用。

烟台植物园以山海城市启动器为规划设计主题，将现状场地东部区域的自然山体与西部区域因汇水形成的冲沟和当地农民自建的两个蓄水水塘统筹考虑，有效提升了城市生态可持续性与区域综合效益。

Plant is the most important element in the garden, and it is also an indispensable natural resource in people's life. According to statistics, there are more than 300000 species of higher plants in the world. In China, there are more than 25000 species of seed plants alone, including about 8000 species of trees and shrubs. People's exploration of the plant world has never stopped, and this kind of exploration has gradually evolved into a collection behavior. People engaged in this activity are often called "plant hunters". It can be said that the process of plant collection is full of unknowns and uncertainties, which also plays an important role in the development of world botany and the construction of botanical garden.

The history of plant collection and the development of botanical garden come down in one continuous line. Many specialized plants were collected and cultivated in many ancient gardens at home and abroad, such as Shanglin garden in the Qin and Han Dynasties, European monasteries in the 5th century and so on. At that time, most of the plants collected were medicinal plants with certain economic value, and only a small part of them were used as ornamental plants.

In the 16th century, Pisa Botanical Garden and Padua Botanical Garden were built in Italy. The function of botanical garden gradually changed from simply collecting and cultivating cash crops to plant introduction, scientific research, and teaching. At this time, the structure of botanical garden continued the regular pattern of medieval monastery garden.

In the 17th century, with the development of plant taxonomy and the impact of Renaissance on culture and economy, a world war of plant plunder began slowly. The botanical garden also opened areas for studying and collecting plant resources from all over the world, and the structure of the botanical garden was more flexible and natural. However, facing the plants from different regions, how to successfully introduce and make the plants survive became a new topic, which also promoted the development of cultivation facility technology. The industrial revolution provided the possibility for this development, with some large tropical greenhouses built. Since then, it has become the primary task of botanical garden to introduce and collect rare plants from all over the world.

Since the 19th century, with the rapid development of urban construction, the demand of public green space for plants with good growth characteristics has increased. Botanical gardens have gradually taken on the task of introduction and domestication, breeding, and trial planting of new varieties of plants. Some botanical gardens have also opened special introduction and domestication areas, making botanical gardens a step forward in supporting industrial development. It is also from this stage that the research-oriented botanical garden emerged in China, and the transformation from the specialized planting area of classical garden to the botanical garden with public scientific research nature was completed. However, compared with the advanced botanical gardens, there was still a big gap.

Ever since the 20th century, people gradually realized that the rapid development and excessive demand for nature are aggravating the destruction of many plant species. It is assumed that at the end of the 21st century, the deterioration of the ecological environment will cause 1/2 of the plants to face the threat of survival, and more than 2/3 of the vascular plants may disappear completely. People gradually realize that the botanical garden has an important role in the protection of rare and endangered plants. The botanical garden is regarded as the "ark" to save plant species from extinction.

Today's botanical garden, while carrying out its own construction, has also set up a sharing mechanism of germplasm resources and built a more comprehensive scientific research exchange platform. So far, more than 3600 botanical gardens have joined the Botanic Gardens Conservation International. Based on ensuring the green space function of general parks, they also have the unique attributes of collecting and recording living plants for protection, exhibition, scientific research, popularization and utilization.

Every city needs a botanical garden. According to the "State Standard for Ecological Garden City", there should be at least one botanical garden with an area of more than 40 hectares in prefecture level cities, which promotes the vigorous development of China's botanical garden industry. The issue of "Botanical Garden Design Standards" in 2020 is of great significance to ensure the design quality and give full play to the comprehensive benefits of botanical garden.

The construction of ecological civilization has brought about a new change in the values of Chinese society. It has become a new development trend to conform to nature, protect nature and utilize nature. The construction and development of botanical garden is not only to achieve a leap in the number of constructions, but also to play a greater role in the balance of geographical distribution, the diversity of plant collection, the integrity of functional types, the uniqueness of scientific research, and the sustainability of exchanges and cooperation. With the reform of ecological civilization system, more attention has been paid to the construction of beautiful China. Botanical gardens will become a vivid practice to strengthen the construction of ecological civilization in the future.

The progress of science and technology and the development of artificial intelligence in the new era also create new opportunities for the construction and development of botanical gardens. This chapter selects four practical projects, based on the specific environmental conditions in today's world, puts forward the construction strategy of botanical gardens, which will have important significance for safeguarding the precious green resources and continuing human civilization.

National Botanical Museum Park Planning explores the innovative plant exhibition mode of "integration of museum and garden, combination of landscape and research", and creates a comprehensive exhibition space with mountains and rivers, animals and plants, buildings, and landscapes, and living creatures and human beings coexisting harmoniously.

Planning and Design for the Zhongyuan Yellow River Botanical Garden, as the first botanical garden to systematically collect and cultivate the plants in the Yellow River Basin, has systematically classified and displayed the plants in the North-South transitional zone in the Central Plains of China, put forward the landscape display strategy of plants, and endowed the botanical garden with multiple participation functions.

Guided by the design concepts of close-to-nature and biodiversity, Shijiazhuang Botanical Garden focuses on displaying the protozoa community in Hebei Province. As a biodiversity protection base, an important germplasm resource bank in North China, and an education base combining popular science education and eco-tourism, Shijiazhuang botanical garden has fully played the exemplary role of a comprehensive botanical garden.

Yantai Botanical Garden takes the starter of mountain and sea city as the planning and design theme, taking the natural mountains in the eastern area of the current site, the gully formed by the catchment in the western area and the two water storage ponds built by local farmers into consideration, which effectively improves the urban ecological sustainability and regional comprehensive benefits.

博自然之物，绽盛世之花
——国家植物博览馆园区总体规划

Boasting Things of Nature, Blooming Flowers of Prosperous Times
— National Botanical Museum Park Planning

完成时间：2019年	Time of completion: 2019
建设地点：云南省昆明市盘龙区茨坝片区	Construction site: Panlong District, Kunming City, Yunnan Province
项目面积：12.45 km²	Project area: 12.45 km²
建设单位：昆明大健康投资管理有限公司	Construction unit: Kunming Great Health Investment Management Co., Ltd
获奖信息：2020年中国风景园林学会科学技术奖规划设计一等奖	Awards: First prize of Planning and Design of science and Technology Award of Chinese Society of Landscape Architecture in 2020
主要设计人员：李雄 郑曦 曹珊 郝培尧 姚朋 董丽 尹豪 等	Project team: LI Xiong, ZHENG Xi, CAO Shan, HAO Peiyao, YAO Peng, DONG Li, YIN Hao, et al

  国家植物博物馆园区位于云南省昆明市盘龙区茨坝片区，规划面积占地12.45km²，规划以立山水之势，博自然之物，塑康养之地，绽盛世之花为愿景，展现了中国生物多样性辉煌成就与大健康产业蓬勃发展的盛世花景图卷。国家植物博物馆园区的建设基于昆明独特的气候条件和全国领先的科研资源等一系列优势，构建了完整新颖的植物展陈体系和完善的游憩科普教育体系，展示了中国生物多样性成就。

  规划提出"馆园一体，景研结合"的植物展陈创新模式，首创由国家植物博物馆（主馆+副馆）和国家植物博物馆园区（植物园）共同构成的国家植物博物馆基本形态。园区的室外展陈体系与国家植物博物馆主副馆园区的室内展陈体系互为补充，构建了总面积301hm²的室外活体植物展陈区。无论在活体植物的展陈面积还是展陈质量上，均达到了世界一流植物园的水准。打造由山水、动植物、建筑和景观相互映衬的，活体生物与人类和谐共存的综合性展示空间。

  The National Botanical Museum Park is located in Ciba Area, Panlong District, Kunming City, Yunnan Province. The planning area covers an area of 12.45 km². The planning vision is to create a place of recreation and wellness, and to create a flowering landscape that will showcase the glorious achievements of China's biodiversity and the flourishing development of the health industry. The construction of the National Botanical Museum Park is based on Kunming's unique climate conditions and a series of advantages such as the country's leading scientific research resources. It has built a complete and novel plant exhibition system and a complete recreational science education system to showcase China's biodiversity achievements.

  The design proposed an innovative plant exhibition model of "integration of museums and gardens, combined with landscape research", and pioneered the basic form of the National Botanical Museum, which is composed of the National Botanical Museum (main museum + subsidiary museum) and the National Botanical Museum Park (Botanical Garden). The park has established an outdoor exhibition system that complements the indoor exhibition system of the main and auxiliary parks of the National Botanical Museum, building an outdoor living plant exhibition area with a total area of 301 hectares. Both the display area and the display quality of living plants have reached the level of world-class botanical gardens. Create a comprehensive exhibition space where living creatures and humans coexist harmoniously with mountains and waters, animals and plants, buildings and landscapes set against each other.

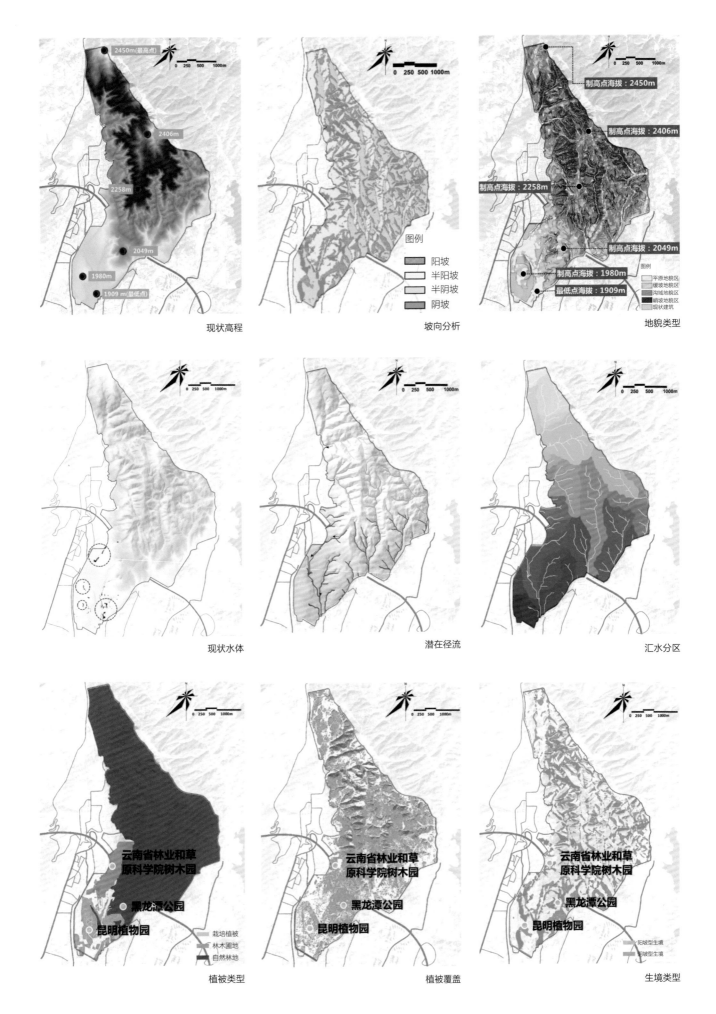

3 植物园 93

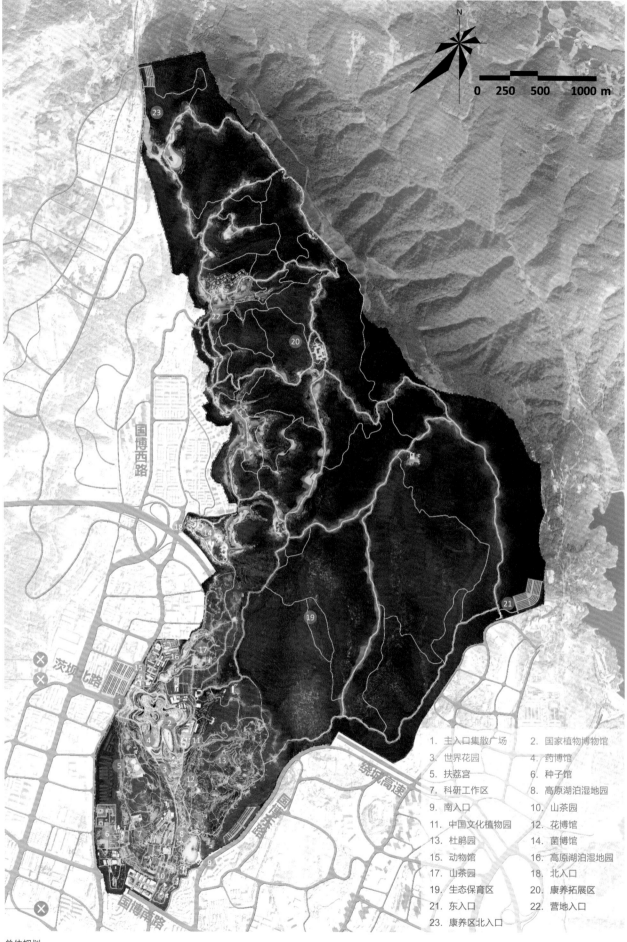

总体规划

1. 主入口集散广场	2. 国家植物博物馆
3. 世界花园	4. 药博馆
5. 扶荔宫	6. 种子馆
7. 科研工作区	8. 高原湖泊湿地园
9. 南入口	10. 山茶园
11. 中国文化植物园	12. 花博馆
13. 杜鹃园	14. 菌博馆
15. 动物馆	16. 高原湖泊湿地园
17. 山茶园	18. 北入口
19. 生态保育区	20. 康养拓展区
21. 东入口	22. 营地入口
23. 康养区北入口	

			富有云南特色的融合性健康产业
世界一流的野生植物种质保存中心和国内顶尖的植物标本中心			
国际知名的植物科学研究中心和国际交流中心			完善的科普教育体系建设
重要的国家战略性植物资源研究开发与保存基地			云南健康生活目的地
		多样化生境营造和功能多元鲜明的科普布局	
	亚热带热带植物的天堂	开展生物多样性保护宣传和国际合作	
	5A级景区标准的最美植物园		
	地区性生物多样性保护的领导者之一		

规划目标1	规划目标2	规划目标3	规划目标4
科学研究与国际交流	植物展陈体系	生物多样性保护	民众互动与社会参与

立山水之势·博自然之物·塑康养之地·绽盛世之花

国家植物博物馆园区的建设填补了国内相关领域的空白。基于昆明独特的气候条件和全国领先的科研资源等一系列优势，密切结合植物科学研究，构建了完整新颖的植物展陈体系和完善的游憩科普教育体系。国家植物博物馆园区的建设不仅是作为植物资源的收集保存、研究开发、科普展示的场所，更能够作为一个地区性生物多样性保护的领导者和展示中国生物多样性成就的窗口。2020年，生物多样性公约第十五次缔约方大会（COP15）将在中国昆明举行，届时，园区将集中向世界展示我国对全球生物多样性保护计划采取的行动和推进植物多样性保护与可持续利用方面所做出的贡献。

园区以规划带动科研发展，通过国家植物博物馆体系的建设，促进场地内科研院所的种质资源收集、活体植物培育和科研科普结合。一是以昆明植物园、树木园和黑龙潭公园的现状活体植物资源为依托，发挥各单位活体植物展陈的特色，以现阶段各院所的科研方向和优质资源为基础，将现有体系整合优化为24个具有国内一流水准的精品专类园。精品专类园整合了以种及品种收集、濒危资源保护见长的昆明植物园内7个类型、23个专类园；以引种驯化、培育保护、经济利用见长的云南省林业和草原科学院树木园内4种类型、33个专类园，以及黑龙潭公园内历史悠久、品类众多的植物专类花园及唐梅、松柏、元杉、明茶等古树名木。

二是以中科院昆明植物研究所与云南省林业和草原科学院的科研强项为依托，结合云南特色生物多样性资源与中国传统植物文化，将首席科学家PI制度作为立园之本，设立了4个具有中国或云南特色，达到世界一流水准的顶级专类园。顶级专类园将现有的科研资源作为构建顶级专类园的基础，通过景观资源协同，实现科研科普的共同发展。

三是利用中科院昆明植物所现有的种质资源库（累计采集保存种子数量10285种82746份，超过中国种子植物总数的三分之一）和标本库（馆藏标本150余万份，是全国第二大植物标本馆）。以国家植物博物馆的建设为契机，提升标准，打造国家战略级野生植物种子保存中心和国内顶尖的植物标本中心。

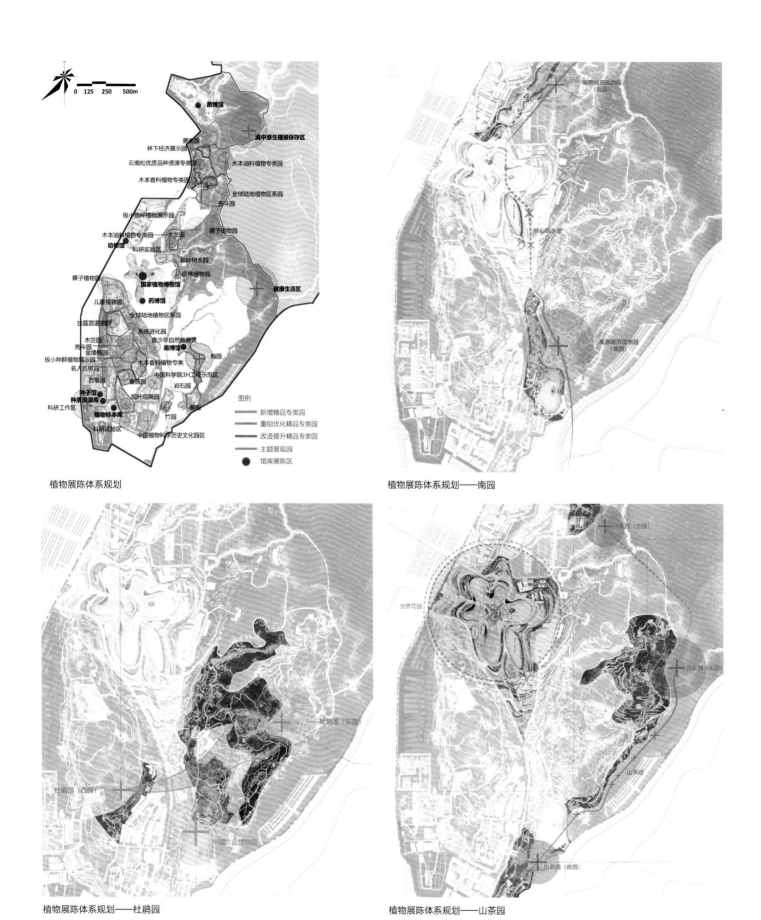

植物展陈体系规划

植物展陈体系规划——南园

植物展陈体系规划——杜鹃园

植物展陈体系规划——山茶园

园区的植物科普展陈体系遵循"馆园一体"的规划理念，形成室内外一体化的科普展陈体系，将以主馆、两库、副馆为主的植物科普展陈场馆、室外展陈环境和动植物生境结合为整体。室外展陈环境由三部分组成，分别为主场馆周边的世界花园、4个顶级专类园、24个精品专类园和8个主题景观区，在植物引种保育和塑造优美景观的同时，结合科研教学与科普教育，展示中国生物多样性成就，传承中国特色植物文化。

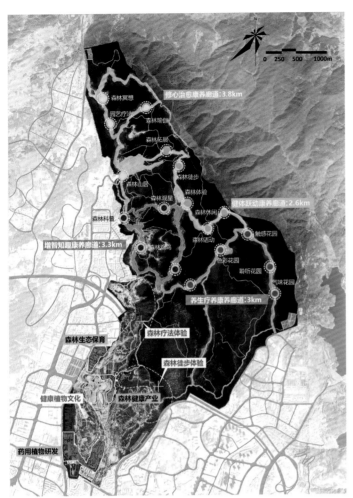

大健康体系规划

大健康体系分析图

园区作为昆明打造"中国健康之城"的示范性项目,利用良好的自然基底,构建了以森林康养为主题的大健康产业示范区。大健康产业示范区一方面以园区的康养功能为基础,融合周边地产、商业、旅游等业态和科研院所的植物康养科研资源,形成大健康产业生态圈。另一方面以产学研用相结合的模式为大健康产业提供支持,引领昆明大健康产业的创新发展,以药物研发、健康保障等领域的科技成果转化工作与中国昆明大健康产业示范区建设紧密结合,更多地向昆明市倾斜科技成果转化项目,带动和支持昆明大健康产业发展。

鸟瞰

黄河流域高质量发展的生态文化新地标
——中原黄河植物园总体规划

A New Ecological and Cultural Landmark of High Quality Development in the Yellow River Basin
— Planning and Design of Zhongyuan Yellow River Botanical Garden

完成时间：2020 年	Time of completion: 2020
建设地点：河南省新乡市原阳县	Construction site: Yuanyang County, Xinxiang City, Henan Province
项目面积：223 hm²	Project area: 223 hm²
建设单位：新乡市人民政府	Construction unit: Xinxiang Municipal People's Government
主要设计人员：李　雄　张云路　等	Project team: LI Xiong, Zhang Yunlu, et al

　　中原黄河植物园作为新乡市沿黄生态带发展规划上的重要节点，是沿黄生态建设的前线工作抓手。长期以来，新乡市与北京林业大学黄河流域生态保护和高质量发展研究院就沿黄生态带发展规划相关工作持续进行着密切合作，中原黄河植物园是双方在沿黄生态带发展规划基础上进行的关键后续项目。

　　中原黄河植物园的规划建设紧密围绕习近平总书记"黄河流域生态保护和高质量发展"战略指示，力求塑造黄河流域人居环境高质量发展的样板，实现黄河文化的保护、传承、弘扬典范，形成黄河流域生态保护和高质量发展在地展示窗口。

　　作为首个系统收集、培育、展示黄河流域植被的植物园，方案提取黄河水系及流域地貌特征，沿园内水系打造黄河五段境域式景观，选取具有典型风貌特征，且适于在新乡露地生长的植物进行展示；在中州平原段重点强调，选取禹贡九州文化作为立意，打造九州万花园，形成全园最大展示集群；同时营建新乡文景苑，选取诗经、牧野、竹贤等新乡本土文化，展示相应主题的特色地域植物。综合以上规划内容，方案形成"黄河一脉、九州同心、新乡画卷"的总体结构。在此基础上结合全园地形，使用克朗奎斯特分类系统，选取太行山、伏牛山、大别山典型植物，对南北过渡带植物进行分类展示。同时为对接周边的高铁、城市和国家种质资源圃，方案将植物园边界打造成为三生融合的主题条带，赋予了植物园多重参与功能。

　　As an important node in the development planning of the ecological belt along the Yellow River in Xinxiang City, Zhongyuan Yellow River Botanical Garden is the front-line work of ecological construction. For a long time, the planning and design team of Institute of Ecological Protection and Quality Development in Yellow River Basin, Beijing Forestry University and Xinxiang government have been working closely together on the ecological belt along the Yellow River. The botanical garden is a key follow-up project on this matter.

　　Zhongyuan Yellow River Botanical Garden is closely related to President Xi Jinping's strategic instructions, striving to create a model for human settlements along Yellow River, to protect the culture, and to set up a model for ecological protection.

　　As the first botanical garden to systematically collect and cultivate the vegetation in the Yellow River Basin, the scheme extracts the water system and geomorphic features of the Yellow River Basin, creates five sections of situational landscape along the water system in the garden, and selects plants with typical features and suitable for growing in Xinxiang; In the section of Zhongzhou, the cultural of Nine Statas is chosen as the intention, and the Nine Statas garden is built on nine islands, which forms the largest landscape vegetation exhibition cluster in the whole park. At the same time, the Culture Garden of Xinxiang is constructed, several local culture of Xinxiang is selected to show the regional plants with corresponding themes. On this basis, the scheme also combined with the whole park terrain, using the Kronquist scientific classification system to classify and display the plants in the North-South transition zone. To connect with the surrounding, the boundary is built into three strip, which endows the park with multiple participation functions.

□ 黄河流域境域式专类园展示脉

1. 青藏高原区
2. 内蒙草原区
3. 黄土丘陵区
4. 新乡文景苑
5. 豫州岛
6. 雍州岛
7. 梁州岛
8. 荆州岛
9. 扬州岛
10. 徐州岛
11. 青州岛
12. 兖州岛
13. 冀州岛
14. 入海三角洲区

□ 三山分类园

1. 太行山分类园
2. 伏牛山分类园
3. 大别山分类园

□ 边界主题园带

1. 生活主题园带
2. 生产主题园带
3. 生态主题园带

□ 建筑及功能区域

1. 黄河生态文明馆
2. 展览温室
3. 园艺体验中心
4. 高山植物馆
5. 科研生产区域
6. 管理服务区域

总平面

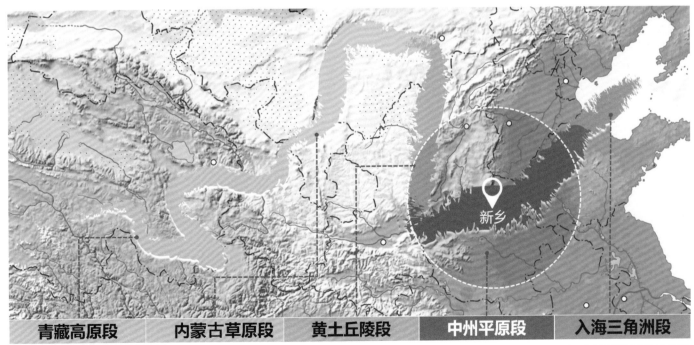

黄河五段：方案对黄河流域景观特征进行艺术提炼，形成青藏高原段、内蒙古草原段、黄土丘陵段、中州平原段、入海三角洲段五段典型境域式景观

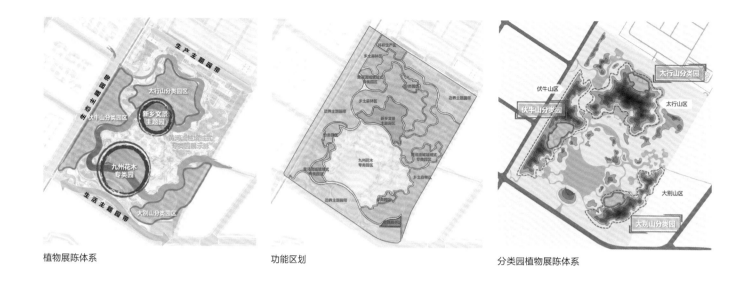

植物展陈体系　　　　　　　　　　功能区划　　　　　　　　　　分类园植物展陈体系

全园整体植物展陈结构为一脉、两心、三区、三带

"一脉"为黄河流域境域式专类园展示脉，依托园内核心水系，形成黄河自然人文为线索的不同境域专类植物展示。"两心"为九州花木专类园和新乡文景主题园，是黄河脉上核心、景观核心与全园最大的展示空间，系统全面地展示新乡自然与人文特色主题。"三区"为太行山分类园区、伏牛山分类园区、大别山分类园区，依托园内三山地形，与过渡带植物区系地域特征融合进行植物科学分类展示。"三带"为生产主题园带、生态主题园带、生活主题园带，依托植物园三条边界，整合周边绿色产业资源，引入游憩功能，形成功能化、景观化的城市开放边界。

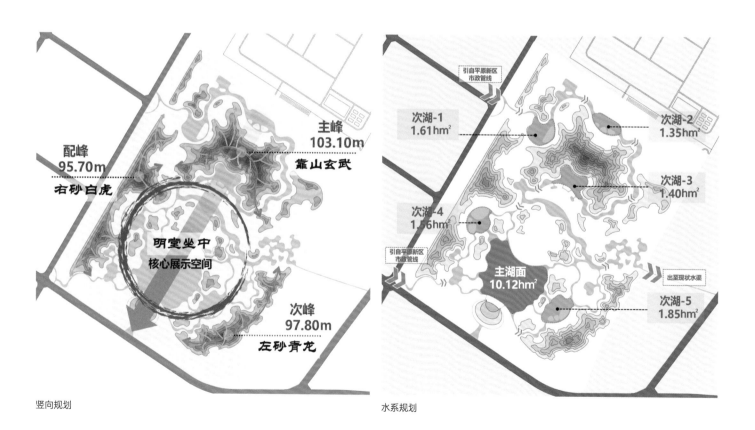

竖向规划　　　　　　　　　　　　　　　　　　水系规划

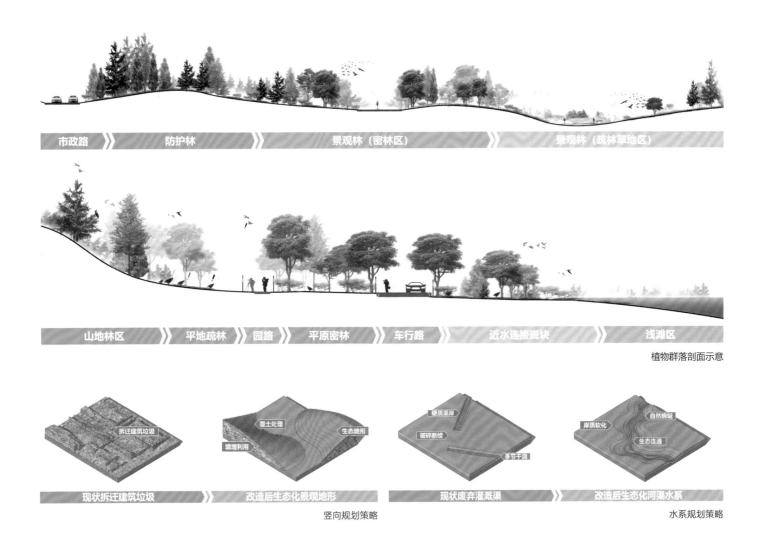

植物群落剖面示意

竖向规划策略　　　　　　　　　　　　　水系规划策略

园区中南部现状为近科楼村，未来村落拆迁后将产生大量建筑垃圾遗留，对这些建筑垃圾进行运输和处理的花费较大。因此，规划提出变废为宝、就地利用的竖向规划策略，对建筑垃圾进行就地夯实处理并覆土填埋利用，节省运输开支的同时塑造生态地形地景，形成将村落拆迁建筑垃圾用于竖向规划的生态转换示范。

园区现状散落遗留有韩董庄灌区三干一支、韩董庄灌区三干二支等若干废旧灌溉渠，渠体硬质，缺乏生态功能，同时破碎断续、亟需整合连通。规划提出沟渠软化、生态串联的水系规划策略，对硬质渠岸进行自然软化，营造多样的动植物群落生境空间，形成将农田废弃灌溉渠进行生态化连通改造的生态修复示范。

边界主题园带效果

近自然城市生物多样性保护的绿色休闲地
——河北石家庄植物园规划设计

A Green Leisure Place of Near-nature Urban Biodiversity Protection
— Planning and Design of Shijiazhuang Botanical Garden in Hebei

建成时间：2003 年	Time of completion: 2003
建设地点：河北省石家庄市鹿泉区	Construction site: Luquan District, Shijiazhuang City, Hebei Province
项目面积：136 hm²	Project area: 136 hm²
建设单位：石家庄市政府	Construction unit: Shijiazhuang Municipal People's Government
获 奖 信 息：2019 年国际风景园林师联合会亚太地区公园及开放空间类荣誉奖	Awards: Award International Federation of Landscape Architects Asia-Pacific Region Award (IFLA AAPME) Parks and Open Space Honorable Mention in 2019
主要设计人员：李 雄 李运远 郑 曦 蔡凌豪 胡文芳 等	Project team: LI Xiong, LI Yunyuan, ZHENG Xi, CAI Linghao, HU Wenfang, et al

　　石家庄植物园位于河北省石家庄市西北部，为进一步发挥植物园科研科普、游览、观赏、休闲娱乐等多项功能，以及保护和改善市区生态环境的作用，植物园进行了扩建。建成后的植物园成为一个以植物展示为主题的公园，也是郊区具有科研、娱乐等功能的绿色生态休闲基地。2006 年石家庄市植物园"五星级公园"标志正式揭牌。作为一个与植物研究推广相关的单位，植物园为河北省植物的收集、研究和展示做出了巨大贡献。截至 2018 年，石家庄植物园汇集了 1136 种植物和数十万棵树木。同时，植物园中具有湿地、森林、草地等多种栖息地。

　　石家庄植物园是一个生物多样性保护基地。该公园结合了近乎自然和生物多样性的设计理念，专注于河北地区的原生生物群落。同时，通过林地、草地、水域等系统的规划，为生物提供了良好的栖息环境，恢复了动植物群落，保护了生物多样性。其次，它也是华北地区重要的种子库。石家庄植物园作为北方地区植物种质资源保护和科学研究的重要基地，不仅栽培和管理现有的园林动植物，而且还积极保护和引进珍贵植物，不断扩大观赏花木的种植面积，丰富了种质资源。最后，它也是科普教育与生态旅游相结合的教育基地。植物园定期与各地区的青年科学技术站合作，开展以科学研究和科普为导向的生态旅游活动成为青少年学习植物的理想场所。

Shijiazhuang Botanical Garden, located in the northwest of Shijiazhuang city, Hebei Province, is a park with the theme of plant exhibition and a Green Eco-Leisure Base in the suburbs with scientific research, entertainment, and other functions. In 2006, Shijiazhuang Botanical Garden won the "Five-Star Park" logo. And as a unit associated with the extension of plant research, the botanical garden has made great contributions to the collection, research and display of plants in Hebei Province. As of 2018, Shijiazhuang Botanical Garden has brought together 1136 species of plants and hundreds of thousands of trees. Meanwhile, the botanical garden has wetland, forest, grassland and other habitats.

The park is a biodiversity conservation base. Combining the design concept of near nature and biological diversity, the park focuses on the protistan community in Hebei. At the same time, this project provides a good habitat environment for organisms, reinstates the flora and fauna community, and protects the biological diversity. Secondly, it is an important seed bank in north China. Shijiazhuang Botanical Garden, as an important base for plant germplasm resources conservation and scientific research in north China, not only cultivates and manages the existing garden animals and plants, but also actively increases the introduction of plants, protects and introduces precious plants, continuously expands the planting of ornamental trees and flowers, and enriches plant germplasm resources. Thirdly, it is also an education base combining popular science education and ecological tourism. The botanical garden regularly cooperates with the youth science and technology stations of various areas to carry out scientific research and science-oriented eco-tourism activities providing an ideal place for young people to learn plants.

① 叠水	
② 温室	
③ 热带植物	
④ 盆景花园	
⑤ 大草坪	
⑥ 多年生花卉园	
⑦ 波澄湖	
⑧ 玫瑰广场	
⑨ 儿童冒险公园	
⑩ 鱼池	
⑪ 药用植物园	
⑫ 荷花塘	
⑬ 植物科技馆	
⑭ 秋园	
⑮ 湖心岛	
⑯ 竹园	
⑰ 樱花园	
⑱ 富氧大道	

总平面　　　　　　栖息地

热带植物温室总建筑面积约6000多平方米，包含热带和亚热带植物77科300余种，5000余株。大厅分为六个植物区：果实植物区、棕榈区、趣味植物区、芳香植物区、椰林和苏铁区

3 植物园

植物园中常常会开展花卉节、国际交流、科普教育等活动，同时也会进行一系列科研活动

药用植物园占地2.46hm²，种植100余种药用植物，旨在弘扬中医药文化，展示中医药植物资源

花钟广场上放置了由各种花装饰而成的17m高的世纪花钟，其设计和构造采用了独特的回声原理，代表一个美好的未来，提醒人们珍惜时间和生命

木化石森林节点

实景鸟瞰

3 植物园 105

山海城市的绿色启动器
——山东烟台植物园规划设计

Green Starter of Mountain and Sea City
—— Planning and Design of Yantai Botanical Garden in Shandong Province

完成时间：2016年	Time of completion: 2016
建设地点：山东省烟台市莱山区	Construction site: Laishan District, Yantai, Shandong Province
项目面积：228 hm²	Project area: 228 hm²
建设单位：烟台市城管局	Construction unit: Yantai City Administration Bureau

获 奖 信 息：2017年教育部优秀工程勘察设计园林景观设计二等奖
　　　　　　2018年国际风景园林师联合会亚非中东地区经济价值类荣誉奖
　　　　　　2019年中国风景园林学会科学技术奖规划设计一等奖

Awards:　Second prize of Landscape Design for Excellent Engineering Survey and Design of Ministry of Education in 2017
　　　　　Award International Federation of Landscape Architects Asia-Africa Middle East Region
　　　　　Award (IFLA AAPME) Economic Value Honorable Mention in 2018
　　　　　First prize of Planning and Design of science and Technology Award of Chinese Society of Landscape Architecture in 2019

主要设计人员：李　雄　郑　曦　蔡凌豪　姚　朋　戈晓宇　郑小东　等
Project team:　LI Xiong, ZHENG Xi, CAI Linghao, YAO Peng, GE Xiaoyu, ZHENG Xiaodong, et al

　　烟台植物园位于山东省烟台市莱山区南部的低山丘陵区。烟台是一座沿海岸线快速发展的带状城市，随着城市化的快速进行，城市总体规划提出了由"沿海城市"向"山海城市"转变的发展目标。植物园选址在南部山区，意在推动该区的发展，是实现山海城市的重要途径。植物园面积共228hm²，现状场地东部区域为自然山体，西部区域有东部山区汇水形成的冲沟和当地农民自建的两个蓄水水塘。园区被一条新建成的南北向省道穿过，分为东西两个部分。规划目标是构建一个以植物展示为特色的公园，西区是主要的植物展示与游人活动区，东区则是山林游览区。通过此项目提升城市生态可持续性，提高区域价值吸引力，促进经济发展。这里作为区域带动的引擎被寄予厚望，但是城市干道穿越基址，切断了山体、冲沟和水库之间的水脉联系，导致现有水塘无法蓄水，同时山区雨水在道路一侧汇集无法排出。现状景观破碎，但是立地条件多样，山林保护较好，为植物园建设提供了较好的场地基础。

　　Yantai Botanical Garden is located in the low hilly area in the south of Laishan District, Yantai City, Shandong Province. Yantai is a belt city with rapid development along the coastline. With the rapid urbanization, the urban master plan puts forward the development goal of transforming from "coastal city" to "mountain and sea city". The botanical garden is located in the southern mountainous area, which is intended to promote the development of the area and is an important way to realize the mountain and sea city. The botanical garden covers a total area of 228 hectares. The eastern part of the current site is a natural mountain, and the western part is a gully formed by the catchment of the eastern mountain area and two water storage ponds built by local farmers. The park is crossed by a new north-south provincial road. The planning goal is to build a park featuring plant exhibition. The western part is the main plant exhibition and visitor activity area, and the eastern part is the mountain forest tourist area. Through this project, we can enhance the sustainability of urban ecology, enhance the attractiveness of regional value and promote economic development. Construction here has been given high hopes as an engine of regional drive, but the newly built urban trunk road passes through the base site, cutting off the water vein connection between the mountain, gully and reservoir, resulting in the existing pond unable to store water, while the rain water in the mountain area can not be discharged on one side of the road. The current landscape is fragmented, the vegetation type is single, and fruit trees are the main ones. However, the site conditions are diverse and the mountain forests are well protected, which provides a good site foundation for the construction of botanical gardens.

鸟瞰效果

❶ 主入口（西入口）
❷ 东入口
❸ 规划展览馆
❹ 温室
❺ 景观湖
❻ 商埠花园
❼ 科研建筑
❽ 科研温室
❾ 专类园展示区
❿ 接待中心
⓫ 观赏植物分类园区
⓬ 水生植物花园
⓭ 知戒名
⓮ 芳菲墅

总平面

3 植物园 107

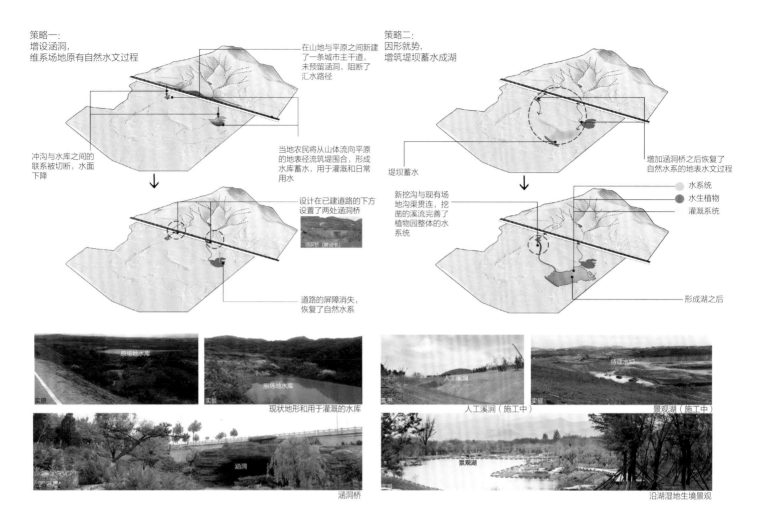

108　理地营境　生态文明建设背景下风景园林实践

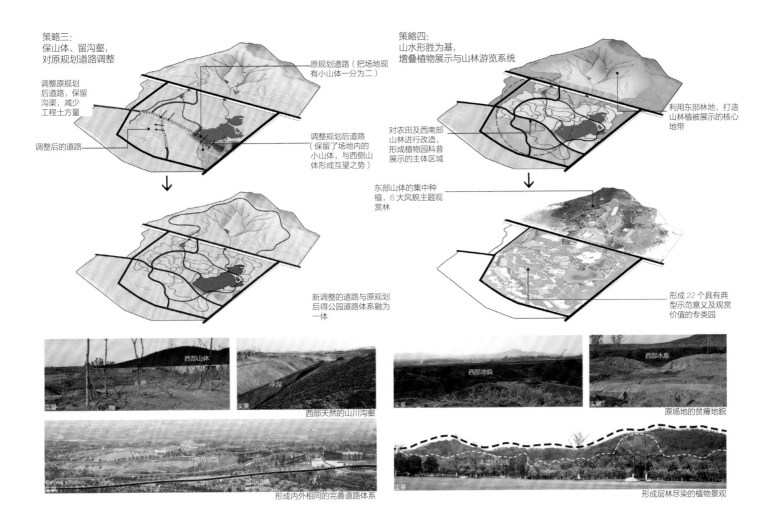

策略三：
保山体、留沟壑，对原规划道路调整

策略四：
山水形胜为基，增叠植物展示与山林游览系统

植物园西园鸟瞰

在原有景观系统的基础上,建立了连贯的水系,隔湖远眺东部自然山林,视线开阔,湖面微波荡漾,近石、远山在碧湖的映衬下相得益彰

低洼区大面积种植了乡土花卉,形成美丽的生态花卉谷

湿地和环湖步道：走在湖泊湿地的休闲步道设施上，游客可以近距离感受自然栖息地带来的乐趣

山林游览区栈道：游客可以穿过树林，在森林中呼吸新鲜空气，与大自然亲密接触，感受山野

玉兰园：种植了乡土树种玉兰的艺术感专类园，玉兰与精致的地形相结合，创造出一个小型专类花园空间

紫薇木槿园：游客可以在这里欣赏到夏季开花的专类植物

主入口区由 8 片从四周向中心逐渐升起的耐候钢板饰面的墙体组成，简约有力，造型现代，极具美感

出入口开敞的草坪将视线引向远处的自然山林，形成视线开阔的景观，草坪、茂密的植物和远山尽收眼底

花岗石错落有致地镶嵌在东入口的草坪上，自然地将坚硬的场地与柔软的草坪相结合

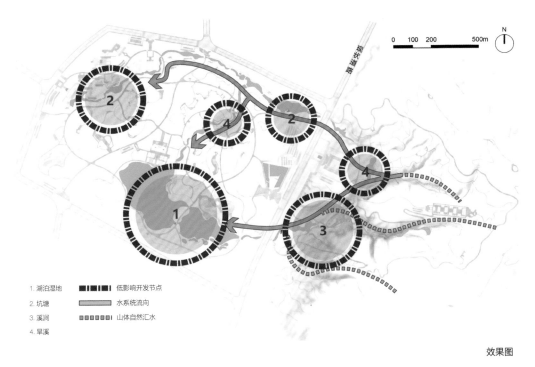

1. 湖泊湿地 ▬▬▬▬ 低影响开发节点
2. 坑塘   ▬▬▬ 水系系统流向
3. 溪涧   ▭▭▭▭ 山体自然汇水
4. 旱溪

湖泊

湿地

效果图

低影响开发

设计中利用了 SWMM 径流模拟软件对方案进行了模拟评估,量化场地的整体雨水径流情况,得出消纳园区雨水径流所需的水体规模。以此为依据,扩大水体面积,形成 3.65hm² 的景观主湖区,蓄水量达 9.11 万 m³,以及植草沟、滞留池、具有净化储存功能的湿地等不同尺度的生物滞留设施,水体总面积为 7.53hm²,蓄水量达 16.38 万 m³。各类型低影响开发的节点,在提供景观价值的同时、营造了水生植物生境,并用于植物园内的灌溉,构建了植物园综合水系。

项目景观绩效评价

烟台植物园建设完成后,选取了生态、社会和经济三个方面共 14 个指标对项目进行了建成前后景观绩效评价,可以看出烟台植物园的建成对当地的生态环境有所增益,提供了丰富的社会功能,同时对区域经济产业产生了带动作用。

坑塘

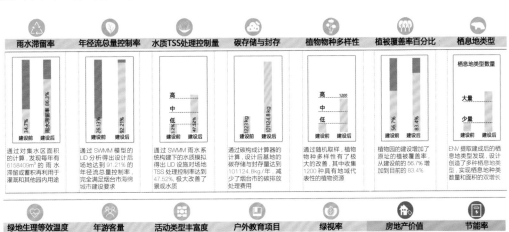

溪涧

旱溪

4

生态修复与郊野公园

Ecological Restoration and Country Parks

近年来，我国高速城镇化和城市扩张建设取得了举世瞩目的成就，同时也在生态环境、基础设施、公共服务、城市文化、城市品质方面存在着许多问题。2015年，在住房和城乡建设部的积极推动下，"城市双修"从三亚发起并进入公众视野，星星之火开始在全国燎原。

城市双修是指"城市修补、生态修复"，用生态的理念，修复城市中被破坏的自然环境和地形地貌，改善生态环境质量；用更新织补的理念，拆除违章建筑，修复城市设施、空间环境、景观风貌，提升城市特色和活力。进行"城市修补、生态修复"是治理"城市病"、保障改善民生的重大举措，是适应经济发展新常态、大力推动供给侧结构性改革的有效途径，也是城市转型发展的重要标志。

至今，住房和城乡建设部已公布了三批城市双修试点城市，试点城市数量已达到58个，在城市生态修复方面强调注重城市与生态的共生关系、注重自然与人的亲近关系，其中，地缘接近城市中心城区的郊野公园，既对接城市又对接乡野，更加对接人的需求，由此逐渐成为大都市生态修复建设的重要抓手。

从20世纪50年代开始，北京、天津、上海等城市先后提出了绿带的建设理念。1958年，受莫斯科规划影响下编制的北京市城市总体规划中提出城市用地中心区与边缘集团之间以及各边缘集团之间应利用成片的大绿带进行隔离，避免城市"摊大饼式"发展。

郊野公园的发源地可追溯至英国，其政府于1966年发表《郊野公园白皮书》(Leisure in the Countryside)，提出建立郊野公园的决议，意在由此降低人们对乡村资源的破坏，为人们享受郊外休闲娱乐提供一个便利而安静的场所。英国郊野公园建设过程中曾出现由于缺乏经济支持而发展衰竭的现象，但之后郊野公园复兴计划的制定推动了郊野公园的成功建设。

中国香港地区在1976年制定了《郊野公园条例》(Country Parks Ordinance)，将40%的土地开辟为郊野公园，以有效保护动植物原生栖息地，提供动植物的庇护场所，使物种自然繁衍，这为日后香港郊野公园的建设管理提供了强有力的法律保障。郊野公园使香港在高强度城市发展和旺盛土地需求的情况下仍然保持了天然的绿色环境，也为香港市民放松身心舒缓压力、休闲娱乐运动健身、到大自然中游赏学习提供了重要场所。

受到香港地区郊野公园的影响和启发，深圳成为内地最早开始探索建设郊野公园的城市之一。自2003年以来，深圳借鉴了香港郊野公园的规划模式，开始规划建设21个郊野公园，总面积约262km^2，占城市总面积的13%，占林地总面积的27%。同时，《深圳市郊野公园规划》的编制工作已经展开。

上海从2003年起对郊野公园的规划建设进行了初步的探究，以郊野公园与生态湿地相结合的形式，逐步改建市内生态基底良好的林地、湿地、森林以及一些城郊公园，虽然目前并没有直接以郊野公园命名，但已具备了郊野公园的雏形，如已建成的滨江森林公园、吴淞炮台湾湿地公园、顾村公园等。

2000年，北京市启动了第一道绿化隔离带的建设；2003年，北京市第二道绿化隔离带建设工程也正式启动；2007年初，北京市按照城市总体规划开始启动"郊野公园环"的建设，提出要形成第一道绿化隔离地区的公园环及景观带、生态保护带；2016年，《北京市城市总体规划（2016-2035年）》中又提出建设二道绿隔郊野公园环，掀开了郊野公园建设的新篇章。根据北京市园林绿化局公布的名单，全市新建郊野公园达到52处，成为我国建设力度最大、建设速度最快、规模最广、数量最多的郊野公园建设城市。

近年来，南京、天津、成都、杭州、广州、东莞、中山、昆明、海口、石家庄等城市也先后开始了郊野公园的规划与建设，为我国生态文明建设作出贡献。郊野公园因其便利的地理区位和优美的自然环境，为满足人们日益增长的游憩需求提供空间，也保护了城郊自然资源，丰富了城市绿地生态结构，解决了城市无序蔓延的问题，逐渐成为城市建设的重要内容之一。

本章选择了3个实践项目，探索不同视角下郊野公园的建设途径，为提升场地原有生态功能、创造更加广阔舒适的绿色休闲空间、促进区域社会经济发展发挥了积极作用。

晋中市百草坡森林植物园项目地处湿陷性黄土山地，设计采用"生态袋"等灵活措施加固不稳定的土壤，重新组织公园内的排水系统，消除周边山体径流对土壤侵蚀的威胁，形成了健康稳定的公园生态结构。

晋城市白马寺沉陷区生态综合整治工程根据煤矸石倾倒区特点，通过水系规划、人工湿地建设、煤矸石回收利用等生态建设手段，完成场地环境改善、景观提升和产业结构引导，实现白马寺沉陷区的绿色转型。

烟台市夹河生态郊野公园构建了一个以应对城市雨洪为核心、促进城市边缘区发展的滨河郊野公园，实现了高效能应对城市雨洪和低成本建设郊野公园的有机结合，带动了烟台市近夹河区域的更新与发展。

In recent years, China's rapid urbanization and urban expansion have made remarkable achievements, but also left a lot of problems in the ecological environment, infrastructure, public services, urban culture, urban quality. In 2015, under the active promotion of the Ministry of Housing and Urban-Rural Development of the People's Republic of China, "urban double repair" was launched from Sanya into the public view, and a single spark began to start a prairie fire in the country.

Urban double repair refers to "urban repair and ecological restoration", which uses the concept of re-ecology to repair the damaged natural environment and topography in the city and improve the quality of ecological environment; uses the concept of renewal and mending to demolish illegal buildings, repair urban facilities, space environment and landscape, and enhance the characteristics and vitality of the city. It is a major measure to control "urban diseases" and ensure the improvement of people's livelihood and is an effective way to adapt to the new normal of economic development and vigorously promote the supply side structural reform. It is also an important symbol of urban transformation and development.

So far, the Ministry of Housing and Urban-rural Development has announced three groups of pilot cities for urban double repair, and the number of pilot cities has reached 58. In the aspect of urban ecological restoration, it emphasizes the symbiotic relationship between city and ecology, and the close relationship between man and nature. Among them, the country park, which is close to the city center, not only connects with the city but also with the countryside, and better meets the people's demands in the construction of ecological civilization, which has gradually become an important starting point for the construction of ecological repair in metropolis.

Since the 1950s, Beijing, Tianjin, Shanghai and other cities have put forward the concept of green belt construction. In 1958, under the influence of Moscow planning, the master plan of Beijing proposed that the central area of urban land should be separated from the marginal groups, and the marginal groups should be separated by large green belts from each other, so as to avoid the "big pie" development of the city.

The birthplace of country parks can be traced back to the United Kingdom. In 1966, the government issued a country park white paper, *Leisure in the Countryside*, which proposed the resolution to establish country parks. The purpose was to reduce people's damage to rural resources and provide a convenient and quiet place for people to enjoy leisure and entertainment in the countryside. In the process of the construction of country parks in Britain, there was a phenomenon of development failure due to the lack of economic support, but then the formulation of the country park rehabilitation plan promoted the successful construction of country parks.

In 1976, Hong Kong enacted the Country Parks Ordinance, which proposed to allocate 40% of its land to country parks to effectively protect the original ecological habitat of animals and plants, provide shelter for animals and plants, and make species naturally reproduce. This provides a strong legal guarantee for the construction and management of the country park in Hong Kong. The country park keeps Hong Kong in a natural green environment under the condition of high-intensity urban development and strong land demand and provides an important place for Hong Kong citizens to get relaxed, have leisure and entertainment, study and enjoy themselves in the nature.

Influenced and inspired by the country parks in Hong Kong, Shenzhen became one of the earliest cities in the Chinese mainland to initiate the construction of country parks. Since 2003, Shenzhen has learned from the planning model of country parks in Hong Kong and started to plan and construct 21 country parks, with a total area of 262 square kilometers, accounting for 13% of the total urban area and 27% of the total woodland area. At the same time, the Compilation of the Planning of Shenzhen Country Parks has been carried out.

Since 2003, Shanghai has made a preliminary exploration on the planning and construction of country parks. Through combination of country parks with ecological wetlands, it has gradually rebuilt the city's ecologically sound woodlands, wetlands, forests, and some suburban parks. Although it has not been directly named after country parks at present, it has the rudiment of country parks, such as the Binjiang Forest Park, the Wusong Paotaiwan Wetland Park, the Gucun Park, etc.

In 2000, Beijing started the construction of the first isolation green belt; in 2003, the construction project of the second isolation green belt was also officially started; in early 2007, Beijing started the construction of "country park ring" according to the urban master plan, and proposed to form the park ring, landscape belt and ecological protection belt of the first isolation green area; in 2016, the Urban Master Plan of Beijing (2016–2035) proposed to start the construction of "country park ring" in the second isolation green belt, which opened a new chapter in the construction of country parks. According to the list released by Beijing Municipal Bureau of Landscape Architecture, 52 new country parks have been built in the city, and Beijing has become the largest, fastest, and most numerous country park construction city in China.

In recent years, Nanjing, Tianjin, Chengdu, Hangzhou, Guangzhou, Dongguan, Zhongshan, Kunming,

Haikou, Shijiazhuang and other cities have also started the planning and construction of country parks, making contributions to the construction of ecological civilization in China. Because of its convenient geographical location and beautiful natural environment, country park provides space to meet people's growing demand for recreation, protects suburban natural resources, enriches the ecological structure of urban green space, solves the problem of urban sprawl, and gradually becomes one of the important contents of urban construction.

This chapter selects three practical projects to explore the construction approaches of country parks from different perspectives, which play a positive role in improving the original ecological function of the site, creating a broader and more comfortable green leisure space, and promoting the regional social and economic development.

The project site of Jinzhong Baicaopo Forest Botanic Garden is a collapsible loess mountain area. The design adopts flexible measures such as "ecological bag" to reinforce the unstable soil, reorganizes the drainage system in the park, and eliminates the threat of surrounding mountain runoff to soil erosion to form a healthy and stable ecological structure of the park.

According to the characteristics of the coal gangue dumping area, the Ecological Comprehensive Improvement Project of Baimasi Subsidence Area in Jincheng, through water system planning, artificial wetland construction, coal gangue recycling and other ecological construction means, completed the site environment improvement, landscape promotion and industrial structure guidance, and realized the green transformation of Baimasi subsidence area.

Jiahe Ecological Country Park in Yantai City built a riverside country park with the core of dealing with urban rain and flood and promoting the development of urban fringe area, which has realized the systematic combination of efficiently solution to urban rain and flood and low-cost construction of country park and has promoted the rejuvenation and development of the area near Jiahe River in Yantai city.

黄土台塬的生态系统再生
——山西晋中市百草坡森林植物园规划设计

Ecosystem Regeneration in Loess Tableland
— Planning and Design of Baicaopo Forest Botanical Garden, Jinzhong City, Shanxi Province

建成时间：2018 年	Time of completion: 2018
建设地点：山西省晋中市榆次区	Construction site: Yuci District, Jinzhong City, Shanxi Province
项目面积：348.16 hm²	Project area: 348.16 hm²
建设单位：晋中市园林局	Construction unit: Jinzhong Garden Bureau

获 奖 信 息：2019 年教育部优秀工程勘察设计园林景观设计三等奖
2019 年英国景观行业协会国家景观奖国际奖
2019 年国际风景园林师联合会亚太地区自然保护类荣誉奖

Awards: Third prize of Landscape Design for Excellent Engineering Survey and Design of Ministry of Education in 2019
British Association of Landscape Industry (BALI) National Landscape Architecture Award, International Award in 2019
Award International Federation of Landscape Architects Asia-Pacific Region
Award (IFLA AAPME) Nature Conservation Excellence Award in 2019

主要设计人员：李 雄 张云路 林辰松 戈晓宇 李方正 李运远 尹 豪 姚 朋 赵 鸣 刘利刚 郑小东 蔡凌豪 孙漪南 肖 遥 胡 楠 葛韵宇 等

Project team: LI Xiong, ZHANG Yunlu, LIN Chensong, GE Xiaoyu, LI Fangzheng, LI Yunyuan, YIN Hao, YAO Peng, ZHAO Ming, LIU Ligang, ZHENG Xiaodong, CAI Linghao, SUN Yinan, XIAO Yao, HU Nan, GE Yunyu, et al

  百草坡森林植物园项目位于山西省晋中市主城区东部，占地总面积348.16hm²，是晋中市面积最大的中心城市绿地，是集晋中地域特色植物收集与植物景观展示、生态休闲旅游、科普教育于一体的地域性植物园。

  项目所在区域为典型的湿陷性黄土山地，整体地势东高西低，地形条件复杂，多断崖、陡坡，场地中部有一条因常年雨水冲刷形成的南北向黄土沟谷。特殊的土壤结构加之常年不合理的围垦，造成了严重的水土流失和植被退化。场地内约70%面积的土地存在水土流失问题，40%面积的土地无植被覆盖。如何以景观化的手段保留场地的地貌特征、维持场地内的水土稳定并为后续设计提供基础是设计团队面临的首要问题。同时，如何恢复场地破损的植被，为各类生物提供生境，如何应对场地内的人工废弃物，以及如何提升场地的景观吸引力等问题都亟待解决。设计团队希望通过公园建设来恢复生态活力。在尽可能保留现有地形的情况下，设计采用"生态袋"等灵活措施加固不稳定的土壤；重新组织公园内的排水系统，并以"分层雨水排水系统"的方式消除周边山体径流对土壤侵蚀的威胁；在保证水土稳定的基础上，以晋中地区周边原生植被群落和场地现有植被为基础，丰富植被群落结构，创造美丽的植物景观，形成了健康稳定的公园生态结构。项目建设以来，恢复了140hm²的绿地面积，总体生境单元饱和指数达到72%，生物多样性提高了81%，消除了390670m³的年雨水径流，解决了存在的土壤侵蚀问题。

  Baicaopo Forest Botanical Garden is located in Jinzhong city in the east, Shanxi Province. The garden covers a total area of 348.16 hectares, is the largest green land at the center of the city, a regional botanical garden with the collection of plant and plant landscape, ecological leisure tourism, and popular science education.

  The area where the garden is located is a typical collapsible loess mountain area with complex terrain, including many cliffs and steep slopes, high in the east and low in the west. In the middle of the site, there is a north-south loess gully formed by perennial rainwater. The special soil structure and unreasonable reclamation caused serious soil and water loss and vegetation degradation problems. About 70% of the land in the land has the problem of soil erosion, and 40% of the land has no vegetation cover. How to preserve the geomorphic features of the site and maintain the soil and water stability to provide the basis for the subsequent design is the primary problem facing by the design team. At the same time, how to restore the damaged vegetation of the site, provide habitat for all kinds of organisms, how to deal with the artificial waste and how to improve the attraction of the site are also urgently to be solved. The design team wants to restore ecological vitality through the construction of the park. While preserving the existing terrain as much as possible, the design uses flexible measures such as "ecological bags" to strengthen the unstable soil, reorganize the drainage system in the park, and eliminate the threat of soil erosion from surrounding mountain runoff by means of a "stratified stormwater drainage system". On the basis of ensuring the stability of soil and water, the park has formed a healthy and stable ecological structure by enriching the vegetation community structure and creating beautiful plant scenery on the basis of the native vegetation community around Jinzhong area and the existing vegetation on the site. Since construction, 140 hectares of acreage have been restored, the overall habitat unit saturation index has reached 72%, biodiversity has increased by 81%, and 390670 m³ of annual stormwater runoff has been eliminated. Solved the existing soil erosion problem.

场地现状1

场地位于山西省晋中市主城区东部，西接晋中主城区东外环，北与规划的安宁街东延线相接，紧邻未来规划的汽车机械园区，南倚石太铁路，总面积约348.16hm²。

自然现状：项目所在区域为典型的黄土地貌，西侧为黄土丘陵台地，东侧为较为开敞的黄土塬。地形条件复杂，多断崖、陡坡，场地中部有一因常年雨水冲刷形成的南北向黄土沟谷。场地存在大面积的裸露荒地，除中部分布有约78hm²的银杏林及集中分布在场地东部的少量林地，其余植被多为长势较差的草地、灌丛。

场地内的黄土地貌类型有黄土塬、黄土台塬、黄土沟谷和黄土梁四种各具特色的地貌类型。黄土塬的顶面平坦宽阔、周边为沟谷切割的黄土高地，呈花瓣状。

人工现状：有少量周边村民自发建设的小型砖结构建筑，多数已废弃。场地内部已有较为健全的三纵六横的道路体系，但多为常年开垦形成的土石路，质量较差。

设计挑战：广泛分布在场地内的黄土是一种遇水便会迅速被破坏的特殊土壤，常规工程措施的可行性有限。因此如何以景观化的手段保留场地的地貌特征并维持场地内的水土稳定为后续设计提供基础是首要问题。同时，如何恢复场地破损的植被，为各类生物提供生境，如何应对场地内的人工废弃物以及如何提升场地的景观吸引力等问题都亟待解决。

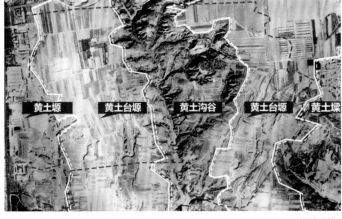

场地现状2

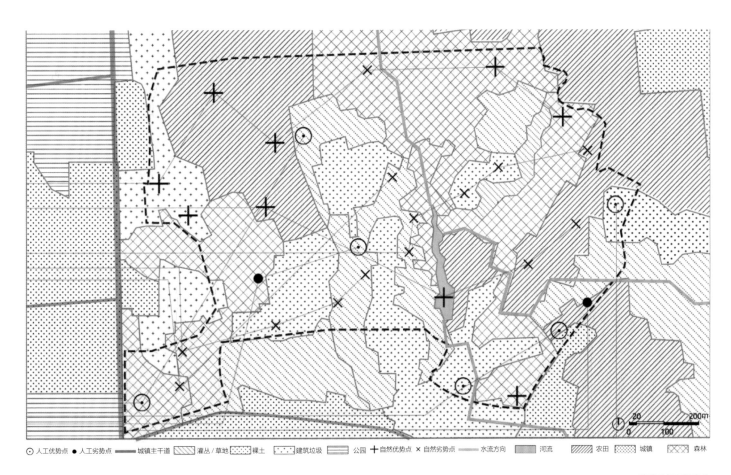

场地现状要素分布

4 生态修复与郊野公园　121

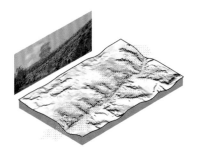  

A 处理极端地形，削减地质灾害风险　　B 梳理场地汇水，构建水安全体系　　C 恢复乡土植被，形成自然基底

通过对现状问题的处理，使场地内不稳定的生态系统得到初步的保护。

设计策略1 修复自然生态

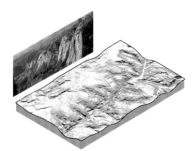 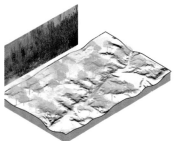 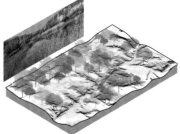 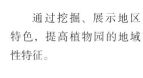

A 保留原有地形特征，展现黄土地质风貌　　B 种植乡土树种，体现晋中自然特色　　C 形成当地适宜生境，丰富全园生态结构

通过挖掘、展示地区特色，提高植物园的地域性特征。

设计策略2 地域风貌打造

 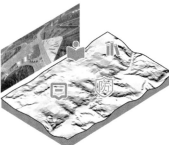 

A 尊重场地空间脉络，形成特色展示体系　　B 植入多样体验内容，完善城市的绿色功能　　C 鼓励公众参与，提升景观认同感

结合多方需求，打造多功能复合的植物园。

设计策略3 多元功能复合

　　从区域生态建设的宏观角度，针对项目所在地黄土地貌特征，基于"生态性""地方性""整体性""功能性""创新性"五大原则提出设计策略。

　　以生态修复为基础，打造具有可持续性的生态植物园；以地域风貌为特色，打造具有地域代表性的乡土植物园；以功能复合为导向，打造具有高度参与性的活力植物园。

现状模式示意

设计方案

场地平面

以生态修复为导向的设计策略

设计尽可能保留现有地形，采用"生态袋"等灵活措施加固不稳定土壤，并通过"分层雨水排水系统"对园区内排水系统进行重组，消除周边山体径流的水土流失威胁。结合晋中地区周边原生植被群落和场地现有植被，丰富群落结构，创造优美的植物景观的同时形成健康稳定的公园生态结构。

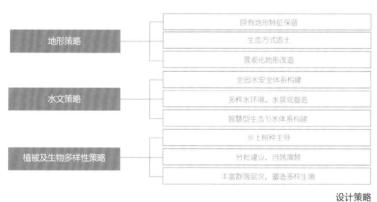

设计策略

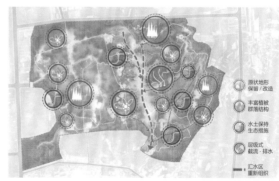

策略落位

实景鸟瞰

4 生态修复与郊野公园

地形策略

生态袋及植被固土模式示意

场地内地形复杂，沟壑纵横，施工设备难以充分介入，并且由于湿陷性黄土的特殊性质，大型工程措施在施工过程中容易破坏已经稳定的土层，造成局部坍塌。设计采用生态袋、植被固土等生态措施以达到稳定地形的目的。

架空的景观栈道

为了尽可能地减少对原地形、植被的扰动，在园中局部场地使用了架空的景观栈道，不仅达到了保持原地形稳定的目的，还完善了全园的交通体系，利用高差为游客带来了丰富的游赏体验。

水文策略

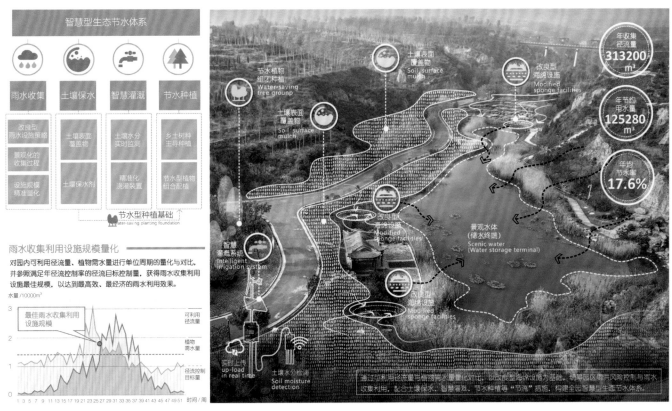

水系统模式示意1

通过可利用径流量与植物灌溉需水量的量化对比，以改良型海绵设施为基础，统筹园区雨洪风险控制与雨水收集利用，配合土壤保水、智慧灌溉、节水种植等"节流"措施，构建全园智慧型生态节水体系。

水系统模式示意2

全园形成了完整的水系，分层截流体系及中部沟谷设计的层级调蓄水体，结合透水铺装、雨水花园、植草沟等措施延缓了两侧山地径流，降低径流峰值流量，维持了区域水安全，降低市政排水压力。全园设置大小、深浅不一的水面若干，形成了丰富多样的水景观、水生境。

植被与生物多样性策略

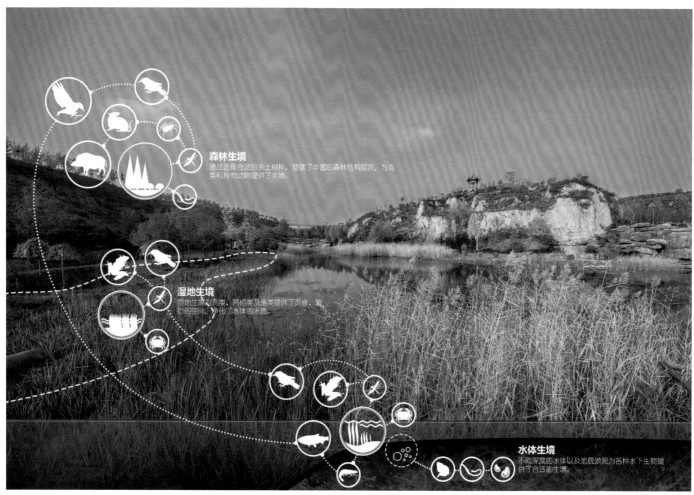

生态系统循环模式示意

设计尽可能保留了场地中原有的植被，并以晋中当地的原生乡土植被群落为依据对场地内的植被进行恢复，创造出了多样的生境类型，塑造了丰富的植物景观，为植物园内形成稳定、可持续的生态结构提供了良好的基底。

蔷薇刺玫园

基于对全园现状植被条件科学、严谨的评估，依据分区、分批次对全园植被恢复进行合理规划。优先对裸地、植被结构单一的区域进行恢复营造景观良好的林地，逐渐形成稳定的植物群落与生态结构。

植物特色展示体系

分类园:依据克朗奎斯特(A. Cronquist)系统展示从低等植物到高等植物的发展进化过程,规划收集华北地区为主的温带植物区系树种,合理配置,形成整体风貌优美、内部变化丰富、各具特色的植物景观。分类园区总面积为 95.5hm^2,共分为 14 个小园,收集种植 91 科 442 种植物。

分类园实景鸟瞰

木兰亚纲分类园

专类园：专类园从不同的植物主题展示出不同的种植方式和布局以及植物景观与建筑、地形的多样组合方式。专类园区占地面积 42.2hm²，包括紫藤花园、花草甸花园、玫瑰园、岩石花园、丁香花园等 9 个专类花园。

蔷薇刺玫园

岩石园

主题园：利用地面天然山谷的中心空间形成八大不同主题的花园，总面积为 26.6hm^2。主题花园同时兼顾水土保持、雨水收集等生态功能。

药用植物园

花谷跌水

煤矸石堆放区的绿色转型
—— 山西晋城白马寺沉陷区生态综合治理工程规划设计

Green Transformation of Coal Gangue Stacking Area
— Ecological Comprehensive Improvement Project of Baimasi Subsidence Area, Jincheng City, Shanxi Province

建成时间：2014 年 10 月	Time of completion: October, 2014
建设地点：山西省晋城市	Construction site: Jicheng City, Shanxi Province
项目面积：108 hm²	Project area: 108 hm²
建设单位：山西省晋城市园林局	Construction unit: Jincheng Landscape Bureau of Shanxi Province

获 奖 信 息：2014 年中国人居环境范例奖
2019 年中国勘察设计协会行业优秀勘察设计奖优秀园林景观设计三等奖
2019 年国际风景园林师联合会亚太地区公园及开放空间类荣誉奖
2019 年教育部优秀工程勘察设计园林景观设计一等奖

Awards: China Habitat Environment Example Prize in 2014
Third prize in Excellent Landscape Design Award of Industry Excellent Survey and Design
Award for China Survey and Design Association in 2019
Award International Federation of Landscape Architects Asia-Pacific Region
Award (IFLA AAPME) Parks and Open Space Honorable Mention in 2019
First prize of Landscape Design for Excellent Engineering Survey and Design of Ministry of Education in 2019

主要设计人员：李 雄 尹 豪 郗大方 郑 曦 刘志成 李冠衡 等
Project team: LI Xiong, YIN Hao, LI Dafang, ZHENG Xi, LIU Zhicheng, LI Guanheng, et al

　　白马寺沉陷区生态综合整治工程位于山西省晋城市，占地 108hm²，是白马寺景区的重要组成部分。项目重点在于寻找煤矸石堆放区的转变方式，解决煤矸石堆放对周边环境和南部市中心生态的威胁，使原场地转型成为一个具有地域特色的植物园，收集展示晋城盆地和周围地区的乡土植物种类。根据煤矸石堆放区特点，通过水系规划、人工湿地建设、煤矸石回收利用等生态建设手段，完成场地生态环境改善、景观提升和产业结构引导，实现白马寺沉陷区发展的绿色转型。

　　项目提出了"三色产业结构转型"策略——场地周边的居民原本主要依赖耕种黄土地为主的"黄色经济"和开采黑色煤炭为主的"黑色经济"，生态整治工程的实施使周边经济产业向着以自然教育和旅游服务为主的"绿色经济"转型。在经济层面，场地用地性质的转变将为居民提供大量新的就业机会，引导周边产业转型；在社会层面，项目通过将特色植物资源与自然教育科普相结合的方式，成为晋城市一个重要的自然教育基地。

Ecological comprehensive improvement project of Baimasi subsidence area is located in Jincheng, Shanxi province, covering an area of 108 hm², serving as an important part of the White Horse Temple area. The project focuses on the transformation mode of coal gangue stacking area, eliminating the ecological threat of coal gangue to surrounding environment and the southern city center, establishing a botanical garden with representative native plants of Jincheng and surrounding ares. Based on the features of coal gangue stacking area, the project improves the site ecological environment and landscape, completes industrial structure guidance, and realizes the green transformation of the subsidence area through ecological methods including water system planning, artificial wetland construction and coal gangue recycling.

The project proposes a three-colour economic transformation strategy. Originally, surrounding residents mainly live on yellow economy of crops cultivation and black economy of coal mining. The implementation of the ecological improvement project transforms the surrounding economic industry into green economy of tourism and nature education. Economically, green transformation provides more employment and morphs local industrial structures. Socially, the project aims to become an important hub for nature science in Jincheng by correlating plant resources with educational courses.

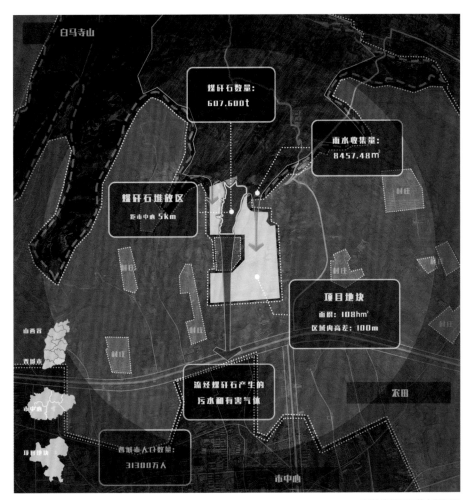

现状汇水分析

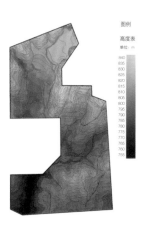

现状竖向分析

区位分析及现状主要问题

场地现状过半为农田，地下为煤炭采空区，时有塌陷和地隙裂缝出现，威胁农业生产和居民生命安全。中部及西北部原为煤矸石堆放区和掩埋场，煤矸石加工生产和多发的自燃现象严重影响城市空气质量；流经煤矸石场区的降水受污染现象严重，威胁农田灌溉和城市水环境安全。当地经济严重依赖煤炭产业，对场地及周边生态环境造成严重破坏；采煤业的无序发展也加剧了当地居民收入的差距，人居环境质量和社会和谐发展受到严重影响。

现状照片

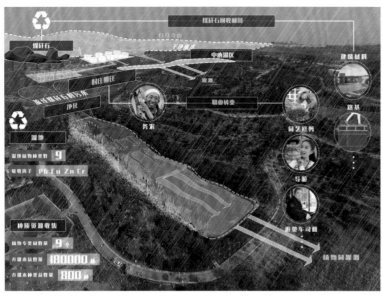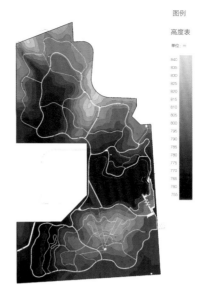

总体策略　　　　　　　　　　　　　　　　　　　　　竖向设计

1 针对煤矸石污染的水处理技术

项目根据水质和流向进行分别处理。针对从山上流下的雨水径流，通过分析现状雨水径流的流向，在场地中部地势低洼的地带设置景观湖泊，收集储存雨水用于植物园内的植物灌溉。针对煤矸石污染的水源在原煤矸石堆放区下游设置生态湿地，选择能够吸附重金属的水生植物，对煤矸石污染径流进行水质净化。

2 煤矸石的转化利用

场地周边的居民原先收入以农业耕种的"黄色经济"和煤矿开采的"黑色经济"为主，这种生产生活方式产值小，同时对生态环境有一定的损害。因此，项目采用了三色产业结构转型策略，针对"黑色经济"剩余的煤矸石，将其制成煤矸石砖和水泥，用于植物园内道路、建筑的建设。

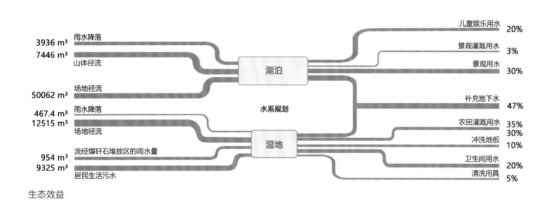

生态效益

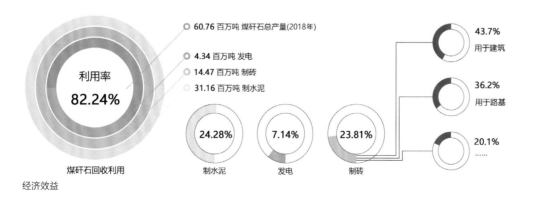

1. 生态修复区	11. 休闲广场
2. 树木园	12. 牡丹园
3. 儿童游乐区	13. 盆景园
4. 观景平台	14. 蔬菜园
5. 蔷薇科专类园区	15. 菊园
6. 景观湖	16. 主入口及停车场
7. 药草园	17. 运动区
8. 紫薇园	18. 办公区域
9. 丁香园	19. 次入口及停车场
10. 生态湿地	

经济效益　　　　　　　　　　　　　　　　　　　　　分区平面

北区规划为生态修复区，主要解决煤矸石带来的淋溶污染和空气污染，植被选择以高覆盖度的灌草为主，可吸纳空气污染的乔木为辅，包括毛白杨、华山松、油松等。

南区规划为植物收集展示和自然教育的区域，主要分区有游乐区、蔷薇科专类园区、蔬菜园、丁香园、紫薇园等，进行种质资源的收集。

生态修复区
儿童游乐区
蔷薇科花园

药草园
牡丹园
菊园
紫薇丁香园
生态湿地

场地总平面

4 生态修复与郊野公园

菊园

药草园

蔬菜园

主入口

景观湖南岸

生态湿地

实景鸟瞰

4 生态修复与郊野公园 135

绿色郊野空间的弹性节约
——山东烟台夹河生态郊野公园规划设计

Elastic Saving of Green Country Space
— Planning and Design of Jiahe Ecological Country Park, Yantai City, Shandong Province

建成时间：2015 年	Time of completion: 2015
建设地点：山东省烟台市芝罘区	Construction site: Zhifu District, Yantai City, Shandong Province
项目面积：31.6 hm²	Project area: 31.6 hm²
建设单位：烟台市莱山区城市管理局	Construction unit: City Administration Bureau of Laishan District, Yantai City

获 奖 信 息：2017 年教育部优秀工程勘察设计园林景观设计二等奖
2019 年英国景观行业协会国家景观奖国际奖
2020 年中国风景园林学会科学技术奖规划设计一等奖
巴塞罗那国际景观双年展罗莎·芭芭拉国际景观入围奖

Awards: Second prize of Landscape Design for Excellent Engineering Survey and Design of Ministry of Education in 2017
British Association of Landscape Industry (BALI) National Landscape Architecture Award, International Award in 2019
First prize of Planning and Design of science and Technology Award of Chinese Society of Landscape Architecture in 2020
Barcelona International Landscape Biennale Rosa Barbara International Landscape Finalists Award

主要设计人员：李 雄 姚 朋 戈晓宇 郑小东 李运远 肖 遥 林辰松 宋 文 李方正 等
Project team: LI Xiong, YAO Ming, GE Xiaoyu, ZHENG Xiaodong, LI Yunyuan, XIAO Yao, LI Chensong, Song Wen, LI Fangzheng, et al

项目位于山东省烟台市外夹河畔，城市南部的边缘地带，西南侧以夹河为界，东、北侧均为新修的市政道路，西侧隔夹河支流与机场高速相接，设计总面积 31.6hm²。项目依托场地现有条件，在满足功能性、景观性的同时，提出针对现状水环境、水安全、水生态、水资源问题的弹性设计策略。

高效能应对城市雨洪：项目秉持"存蓄、消纳、迟滞"的设计原则，将夹河郊野公园作为城市面向夹河的径流防线，同时，也形成面向城市的弹性缓冲区域。选择利用植物、土壤等自然材料，石笼、生物滞留池等工程措施，对地表径流进行净化处理。

低成本建设郊野公园：项目基于弹性理念，通过引入低维护乡土树种，丰富植物群落类型，提高生态护岸比例；公园突出郊野特征，对现状植被和农田进行梳理改造，在全园形成健康步行道系统和湿地栈道科普系统。

项目于 2015 年建设完成，构建了一个以应对城市雨洪为核心、促进城市边缘区发展的滨河郊野公园，实现了高效能应对城市雨洪和低成本建设郊野公园的有机结合。

The project is located on the bank of the Waijiahe River in Yantai City, Shandong Province, the southern edge of the city, with a total design area of 31.6 hectares. Relying on the existing conditions of the site, the project puts forward flexible design strategies for the current water environment, water security, water ecology and water resources problems while meeting the requirements of functionality and landscape.

Efficient response to urban rain and flood. The project adheres to the design principle of "storage, consumption and lag", and regards Jiahe Country Park as the city's runoff defense line facing Jiahe River, and Jiahe River as an elastic buffer area facing the city. Select the use of plants, soil and other natural materials, stone cages, biological retention ponds and other design elements to purify the surface runoff.

Low-cost construction of country parks. Based on the concept of elasticity, the project creates a variety of near-natural habitats and animal habitats by introducing low-maintenance native tree species, plant landscape level. The park highlights the characteristics of the countryside, combs and rebuilds the current vegetation and farmland, forms a healthy walkway system and a wetland science plank system in the whole park.

The construction of the project was completed in 2015, building a riverside country park with the core of dealing with urban rain and flood as the core and promoting the development of urban fringe, realizing the organic combination of high-efficiency response to urban rain and flood and low-cost construction of country parks.

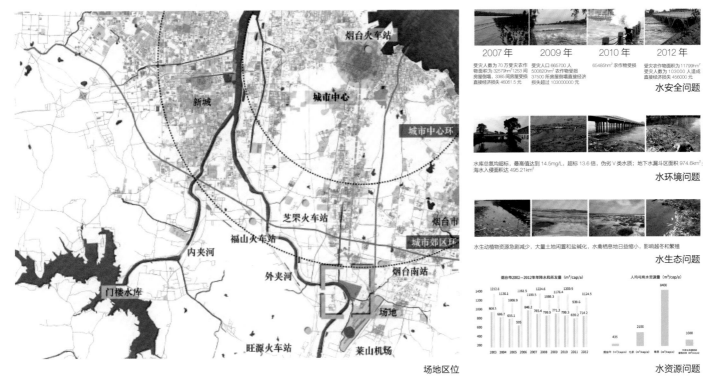

场地区位

水安全问题

水环境问题

水生态问题

水资源问题

夹河是烟台市的母亲河，由内夹河和外夹河合流而成，发源于海阳市郭城镇牧牛山，自南向北流经海阳、栖霞、牟平、福山、莱山、芝罘及经济技术开发区等七市区。流域面积 1072km²，河长 65km，干流比为 0.00132，地貌以低山丘陵为主，地势呈西南高、东北低，流域上游宽，下游窄，呈"梨"型。流域平均长度 60km，宽度 15.5km。项目距市中心 14km，邻近莱山机场，属于城市南部边缘地带。

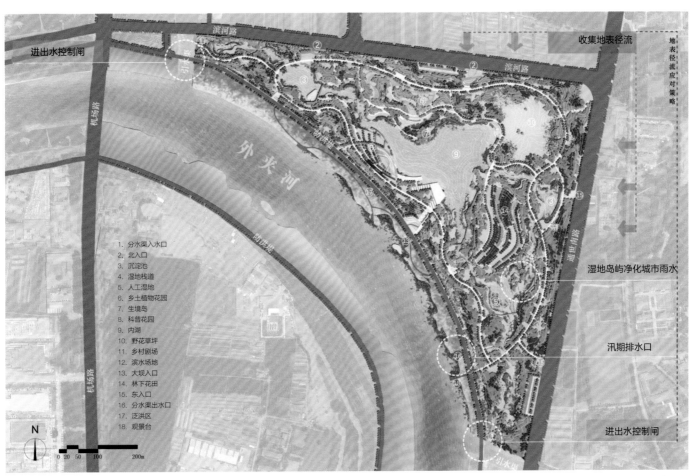

1. 分水渠入水口
2. 北入口
3. 沉淀池
4. 湿地栈道
5. 人工湿地
6. 乡土植物花园
7. 生境岛
8. 科普花园
9. 内湖
10. 野花草坪
11. 乡村剧场
12. 滨水场地
13. 大坝入口
14. 林下花田
15. 东入口
16. 分水渠出水口
17. 泛洪区
18. 观景台

总平面

•水资源	•水环境	•水安全	•水生态
9.84万m² 城市径流收集量	92% 河道改造	70.7% 径流洪峰降低量	416.8kg PM10去除量
8.85万m² 节约用水量	92%（径流）73%（河水） 总固体悬浮物去除量	12.75h 径流洪峰推迟时间	190.1kg NO₂去除量
10.06% 地下水补充率	91%（径流）47%（河水） 总磷去除量	2016—2017年五年一遇，1小时雨量测量值	303.7t CO₂吸收量
2017年测量值	94%（径流）22%（河水） 总氮去除量		211.7t O₂释放量
	2016—2017年两年一遇，1小时雨量测量值		2018年测量值

景观绩效

鸟瞰

专类花园展示了乡土花卉种类，条石坐凳与砾石铺装表现出了典型的郊野公园特征，同时也为植物科普活动提供了良好的活动场所

专类花园中的弧形锈铁板与花卉相互衬托，科普属性和休闲属性在小场地中适当结合

郊野公园的活动场所大多隐藏在地形围合的树林中，有着不同于城市公园的朴野特征，人们在此休闲放松，成为家庭共度周末的完美场所

野花沿着慢跑道和场地自然生长，为郊野公园增添了许多野趣

砾石路面促进雨水的渗透，独特设计的木座椅鼓励公众与内湖直接接触，为动物爱好者观察鸟禽提供场所，同时也是观赏湖景的绝佳场所

石笼生态护岸和沿水设置的木栈道为游人开展滨河游憩提供了舒适野趣的滨水空间，低维护植物的自然生长营造出野趣的氛围

通过物理过滤与生物净化结合的方法为园区的水质提供了保障，使内湖成为水禽的天堂。远处的山林与水中嬉戏的野鸭形成一幅生动的画面

堤坝是郊野公园连接夹河河滨路的入口，采用当地石材砌筑而成，台阶处理高差。雨后场地内没有显著积水，说明地表径流能穿透可渗透地表自由下渗

超过40种低维护、多年生乡土草花生长繁盛，在形成空间边界的同时也提供了广阔的视野

场地采用石材和草镶嵌而成，与周边的野花相映成趣，为人们提供休憩空间的同时增加了野趣体验

雨中湖景

4 生态修复与郊野公园

5
海绵城市与滨水空间
Sponge City and
Waterfront Space

习近平总书记在 2013 年中央城镇化工作会议上明确指出：解决城市缺水问题，必须顺应自然，要优先考虑把有限的雨水留下来，优先考虑更多利用自然力量排水，建设自然积存、自然渗透、自然净化的海绵城市。2014 年 10 月，住建部制定《海绵城市建设技术指南——低影响开发雨水系统构建（试行）》，开启了在结合外国实践经验与本国国情的基础上对城市健康发展模式的积极探讨。2019 年 11 月，习近平总书记视察上海杨浦区滨江开放空间，并具体考察了其城市环境综合治理和海绵城市建设等情况。

海绵城市体系以城市绿地系统为重要载体，同时海绵城市体系的规划结果又对城市绿地系统提出更全面的要求，两者虽然在规划过程中可相互指导，但是在结构布局上仍存在差异。现阶段，在积极探索海绵城市体系构建的浪潮之中，保持对海绵城市体系的理性认知既是平衡城市建设与自然水文相互适应发展的基本前提，也是避免海绵城市体系盲目运动式发展的充分条件，同时有利于促进城市人居环境持续健康发展，有助于海绵城市与城市绿地形成彼此促进、共同发展的状态。

以城市绿地系统为主要载体构建海绵城市体系应关注两者间的耦合关系。在海绵城市体系指导下的城市绿地系统不仅可以满足城市绿地的生态防护、游憩娱乐、文化教育、环境美化等基本功能，同时还可以有效地辅助城市排水设施处理城市雨洪问题，发挥出城市绿地系统更大的潜力。城市绿地系统的构建也会反过来影响海绵城市体系的规划，使海绵城市体系的结构布局更为合理、构建过程更为流畅、作用效果更为明显。

随着海绵城市建设理念的提出，城市绿地开始承担"城市海绵"的功能。绿地除了要满足游憩、生态、景观、防灾避险等基本功能外，还须具备消纳雨水径流的能力。风景园林视角下的海绵绿地指在保证绿地基本功能的前提下，因地制宜地安排低影响开发设施及技术，确保绿地在一定的降雨重现期条件下能基本消纳自身的径流，并有可能收集外部一定范围雨水径流的绿地类型。根据雨水收集的来源可以将绿地分为内源径流型海绵绿地和外源径流型海绵绿地。内源径流型海绵绿地指收集绿地范围内的雨水并对其进行利用的绿地类型；外源径流型海绵绿地强调绿地除了消纳内部的雨水径流外，还能适当接纳周边一定范围内的雨水径流，起到调蓄城市雨洪的作用。在设计中，城市绿地集雨功能的确定是至关重要的。我们一直提倡以问题为导向来确定绿地的集雨功能，内涝风险、水体污染、地下水位降低、水资源短缺等问题都是城市中普遍存在的问题，"渗、滞、蓄、净、用、排"是海绵城市建设中的多样功能类型，以问题为导向的思路能够有效地将问题与功能相对应，将绿地的集雨功能充分发挥，最大化地缓解城市水问题。

在本章节的规划设计项目中，规划设计团队强调绿地在海绵城市中的功能定位的确定，通过 1~2 种功能类型实现绿地的集雨功能和净化功能，达到减排控污的目标。本章节展示的规划设计项目，除了迁安市滨湖东路东侧绿化带景观工程项目外，均完成于"海绵城市"提出之前。

炎帝公园项目位于山西省高平市丹河岸边，项目设计中将绿地的海绵功能定位为"渗"和"净"，通过下渗减排，缓解丹河的排水压力；通过净化控污，减少流入丹河的污染物。

迁安佛教山公园项目是规划设计团队对于海绵城市类项目的第一次尝试，由于场地内部有天然低洼地，设计中将山体径流控制体系与雨水花园相联系，形成了"渗""滞"的功能类型，通过滞留控制径流流速、缓解山体水土流失，通过渗透形成末端的弹性集雨体系，就地消纳场地内部径流。

晋商公园项目位于山西省晋中市，是半湿润地区中降雨量较少的一个城市，规划设计团队将公园的集雨功能定位为"蓄""用"，将内部径流收集排入景观水面蓄水，将城市中水引入公园的水系中来补充景观和灌溉用水。

迁安市滨湖东路东侧绿化带景观工程项目是迁安市成为海绵城市试点城市以后的第一个公园绿地项目，为了实现城市的雨洪安全，这个公园承载着消减周边城市径流的任务。我们将这个公园的集雨功能确定为"蓄用"+"渗滞"，以蓄用来应对常规降雨，补充公园的灌溉用水；以渗滞来应对极端降雨，形成集雨功能的弹性。

安澜湖水公园是以水文化为主要特色的公园，同时具有缓解城市排洪压力的调蓄功能。在设计中将集雨功能定位为"滞""用"，充分发挥大水面的调蓄功能，并依托丰富的水系变化，形成多样的景观和栖息生境。

In the Central Urbanization Work Conference in 2013, President Xi Jinping clearly pointed out that to solve the urban water shortage problem, we must conform to nature, give priority to keeping the limited rainwater, give priority to making more use of natural forces to drain water, and build sponge cities with natural accumulation, natural penetration, and natural purification. In October 2014, the Ministry of Housing and Urban-Rural Development formulated the Technical Guide for Sponge City Construction — Construction of Low-Impact Development Rainwater System (Trial), which initiated a positive discussion on the healthy urban development model based on the practical experience of foreign countries and China's national conditions. In November 2019, President Xi Jinping inspected the riverside open space in Yangpu District, Shanghai, and specifically investigated its comprehensive urban environment management and sponge city construction.

The sponge city system takes the urban green space system as the important. Meanwhile, the planning results of the sponge city system put forward more comprehensive requirements for the urban green space system, although both in the planning process can guide each other, but there are still differences in structure layout. At present, in the tide of actively building sponge city system, maintaining rational cognition of the sponge city system is not only the basic premise of balancing the adaptive development of urban construction and natural hydrology, but also the sufficient condition to avoid the blind development of the sponge city system. At the same time, it is conducive to promoting the sustainable and healthy development of the urban living environment and is also helpful for sponge city and urban green space to form a state of mutual promotion and common development.

To construct sponge city system with urban green space system as the main carrier, we should pay attention to the coupling relationship between them. Under the guidance of sponge city system, urban green space system can not only meet the basic functions of urban green space, such as ecological protection, recreation and entertainment, culture and education, and environmental beautification, but also effectively assist urban drainage facilities to deal with urban rainwater problems and give full play to the greater potential of urban green space system. The construction of urban green space system will also affect the planning of sponge city system in turn, making the structure layout of sponge city system more reasonable, the construction process smoother, and the effect more obvious.

With the proposal of the concept of sponge city construction, urban green space begins to assume the function of "urban sponge". In addition to meeting the basic functions of recreation, ecology, landscape, disaster prevention and avoidance, green space must also absorb rainwater runoff. Sponge green space from the perspective of landscape architecture refers to the type of green space where low impact development facilities and technologies are arranged according to local conditions on the premise of ensuring the basic functions of green space, to ensure that the green space can absorb its own runoff basically under a certain rainfall return period and may collect rainwater runoff in a certain range of external areas. According to the source of rainwater collection, it can be divided into endogenous runoff type sponge green space and exogenous runoff type sponge green space. Endogenous runoff sponge green space refers to the type of green space that collects and utilizes rainwater within the scope of green space. The exogenous runoff type sponge green space emphasizes that the green space not only absorbs the internal rainwater runoff, but also appropriately accepts the rainwater runoff within a certain range of surrounding areas to play the role of regulating and storing urban rainwater. In the design, it is particularly important to determine the rainwater collecting function of urban green space. We always advocate a problem-oriented approach to determining the rainwater harvesting function of green space, waterlogging, water pollution, reduced underground water level, water shortages are common problems in the city, and "infiltration, stagnation, storage, purification, utilization and drainage" are the diverse functional types of sponge city construction. Problem-oriented approach can effectively match problems with functions, give full play to the rainwater collection function of green space, and alleviate urban water problems to the maximum extent.

In the planning and design projects in this chapter, the planning and design teams emphasize the determination of the functional positioning of green space in sponge city, and realize the rainwater collecting and purifying functions of green space through 1-2 functional types to achieve the goal of emission reduction and pollution control. The planning and design projects shown in this chapter, except the landscape project of the green belt on the east side of the Binhu East Road of the Qian'an City, were all completed prior to the concept of "sponge city".

Yandi Park is located on the bank of Dan River in the Gaoping City, Shanxi Province. In the project design, the sponge function of the green space is positioned as "infiltration" and "purification", through infiltration and emission reduction, relieve the drainage pressure of the Dan river, and reduce the inflow of pollutants into the Dan River through decontamination and control.

The Qian'an Buddhist Mountain Park is the planning and design team's first attempt at a sponge city project. Because the site has natural hollows, the mountain runoff control system is connected to the rain garden in the design, forming a functional type of "infiltration" and "stagnation", the flow rate of runoff is controlled by retention, and the soil erosion of the mountain is alleviated. The elastic rainwater collecting system at the end is formed by infiltration, and the internal runoff of the site is absorbed locally.

The Jinshang Park is located in the Jinzhong City, Shanxi Province, which is a city with less rainfall in the semi-humid area. The planning team position the rainwater collection function of the park as "storage" and "utilization". The internal runoff is collected and drained into the landscape water surface to store water, and the urban water is introduced into the water system of the park to supplement the landscape and irrigation water.

The green belt landscape project on the east side of the Binhu East Road of Qian'an is the first park green space project after Qian'an became a pilot city of sponge city. To realize the city's rainwater safety, this park bears the task of reducing the runoff of surrounding cities. The rainwater collection function of the park is defined as "storage and utilization" + "infiltration and stagnation". "Storage and utilization" cope with the regular rainfall and supplement the park's irrigation water; "infiltration stagnation" deal with the extreme rainfall to form the elasticity of the rain-collecting function.

The Anlan Lake Park is a park with water culture as its main feature and has the function of regulating and storing water to relieve the pressure of urban flood discharge. In the design, the rainwater collection function is positioned as "stagnation" and "utilization" to give full play to the regulatory and storage function of large water surface, and to form a variety of landscapes and habitats depending on the diversified changes of water system.

城市与自然之间的集雨绿地
——河北迁安滨湖东路东侧绿带规划设计

Rainwater Harvesting Greenbelt between City and Nature
——Rainwater Collection Green Space between the City and Nature, Qian'an City, Hebei Province

建成时间：2017年		Time of completion: 2017	
建设地点：河北省迁安市		Construction site: Qian'an City, Hebei Province	
项目面积：24.44 hm²		Project area: 24.44 hm²	
建设单位：迁安市园林绿化管理局		Construction unit: Qian'an Landscaping Administration Bureau	

获 奖 信 息：2018年国际风景园林师联合会亚非中东地区雨洪管理类杰出奖
2019年中国风景园林学会科学技术奖规划设计一等奖
2019年教育部优秀工程勘察设计园林景观设计二等奖
2019年英国景观行业协会国家景观奖国际奖

Awards: Award International Federation of Landscape Architects Asia-Africa Middle East Region
Award (IFLA AAPME) Flood and Water Management Outstanding Award in 2018
First prize of Planning and Design of science and Technology Award of Chinese Society of Landscape Architecture in 2019
Second prize of Landscape Design for Excellent Engineering Survey and Design of Ministry of Education in 2019
British Association of Landscape Industry (BALI) National Landscape Architecture Award, International Award in 2019

主要设计人员：李　雄　戈晓宇　林辰松　葛韵宇　邵　明　等
Project team: LI Xiong, GE Xiaoyu, LIN Chensong, GE Yunyu, SHAO Ming, et al

　　2015年迁安市被确立为第一批试点海绵城市，也是京津冀地区唯一一个试点海绵城市。本项目作为迁安市成为首批海绵城市试点后的第一个绿地建设项目，在解决周边地区雨水径流管理的问题上发挥着重要的作用。

　　项目场地位于城市的核心区与城市的中央河道之间。按照上位规划要求，场地需消纳268.8hm²周边城区和道路的雨水径流，雨洪问题是场地的核心问题。此外，如何在解决雨洪问题的同时满足上位规划中对慢行体系的要求和周边居民对户外休闲游憩生活的需求也是设计的重点与难点。

　　设计团队在城市快速发展背景下平衡城市绿地功能和海绵城市建设，提出了蓝色集雨带、红色活力线和绿色游憩面三色共融的设计理念，塑造了富有弹性的多功能空间。该项目的价值不仅在于解决场地周边的雨水径流问题，还在于解决场地缺乏市政道路系统和缺乏活力的问题，将塑造绿色公共空间的最终落脚点放在"人"上，处处以人为本，为人们提供了更加美好的绿色空间体验，成功平衡了绿地与市民需求之间的关系。

In 2015, Qian'an was established as the first batch of sponge city pilot cities, which is also the only sponge city pilot city in Beijing, Tianjin and Hebei. This project is the first green space construction project after Qian'an became the first pilot sponge city, and it plays an important role in solving the problem of stormwater runoff management in the surrounding area. The project site is located between the core area of the city and the central river of the city. According to the upper planning requirements, the site needs to absorb 268.8 hm² of stormwater runoff from the surrounding urban areas and roads, and the stormwater problem is the core issue of the site. In addition, how to solve the stormwater problem and at the same time meet the requirements of the upper plan for the slow walking system and the demand of the surrounding residents for outdoor recreation and leisure life is also the focus and difficulty of the design. In the context of rapid urban development, the design team balanced the function of urban green space and sponge city construction, and proposed the design concept of integrating the blue rainwater collection zone, red vitality line and green recreation surface to create a resilient multi-functional space. The value of the project is not only to solve the problem of stormwater runoff around the site, but also to solve the problem of lack of municipal road system and lack of vitality, and to put the final point of shaping green public space on "people", which is people-oriented in all aspects, providing people with a better green space experience and successfully balancing the relationship between green space and public The relationship between green space and citizens' needs is successfully balanced.

设计场地位于河北省迁安市，曾经是中心城区的一块废弃地，面积24.44hm²，东边是一个高密度住宅区。

迁安属于典型的半湿润气候，夏季集中的降雨极易形成高强度的暴雨，城市面临日趋严重的内涝、暴雨威胁以及水资源短缺问题。因此，场地周围的雨水管理是项目必须面对的基本问题。周边268.8hm²的城市径流会排放进入场地，同时带来大量的面源污染。场地面临着水安全、水资源、水环境三个层面的问题。此外，由于场地西侧的城市道路缺少专属非机动车道，使得场地的可行性降低。长期的废弃也使得地区吸引力降低，周边居民休闲开放空间不足的问题也亟待解决。

如何构建"海绵城市建设背景"下的低影响开发体系；如何在大规模的集雨要求下塑造绿色游憩空间；如何保证公园景观品质符合上位规划要求，成为设计团队所面临的问题。设计团队提出"打造一条融入'海绵城市'建设理念的生态集雨型绿道"的基本设计概念。在场地中置入慢行系统，为周边居民提供休闲游憩场地，建设解决径流问题的雨洪管理系统，共同构成了本次的设计方案。解读项目方案，可以分为"蓝色集雨带""红色活力线""绿色游憩面"三个层面。

场地面临的雨洪问题

蓝色集雨带

水安全——雨水渗透：场地利用原有的天然级配砂石土壤，设计了一条由雨水花园、下沉式绿地等雨水渗透设施构成连续的贯穿场地南北的特色花卉种植带，消减径流的同时强化了植物景观效果。

水资源——雨水收集：场地内部设置了6个蓄水池和4处储水模块进行雨水收集，能同时储存5034m³径流，收集的径流可用于植物灌溉。

水环境——雨水净化：场地内部设置了4处台地式的碎石植被床，径流进入场地前会经过层层净化。

水景观——雨水景观：设计师采用耐候钢和装满建筑垃圾的石笼处理场地内部的高差问题，形成了坚固、统一和富有丰富肌理的设计要素。

碎石植被床：在场地北端设置了一处碎石植被床，通过层台式植被和土壤的净化，市政道路径流的污染问题得到了解决

场地分析

场地分析

台层花园是通过错层的钢板种植池净化西侧道路径流的重要手段，径流通过层层渗透跌落进入底层的种植带，边渗透边流向场地南部。

耐候钢板是迁安本地生产的材料，具有良好的气候适应性，也能形成优质的景观效果，适合应用于雨水设施的景观化处理。

台层花园

暴雨花园

暴雨花园位于场地最南端最低点，其内种植了较耐水湿的展示植物，是低影响开发系统中的末端调蓄区，径流通过层层的渗透和净化最终汇集到这里。此外，当降雨超过设计标准时，所有的径流终会滞留在这里，为城市雨洪分担压力。

场地分析

骑行　　　　　　　　　　　　　　　　　　　　　　　　　　　　跑步健身

红色活力线

　　设计通过建设并行、分行、高低变化的人行道和自行车道来构建慢行体系。其次，慢行体系也用于连接堤、岛形态的景观游憩空间，充分满足周边居民户外游憩的需求。此外，慢行系统主要由透水沥青和透水砖两种材料构成，在雨中能保证基本的使用功能。

　　慢行体系结合周围种植为民众提供更安全、舒适的步行和骑行体验。优美的环境和便捷的慢行交通体系吸引了居民在此开展骑行、跑步健身、散步、遛狗、出游等各项活动。

步行

绿色游憩面

　　场地 80% 以上的面积被植物所覆盖，为市民的游憩活动提供了绿色空间。为了让场地重现活力，设计为各个年龄段的居民设置了各类活动场所，解决了周边高密度居住区缺少户外空间的问题。儿童活动区位于场地南段，与居住用地相邻。设计内容包括互动水池、攀爬设施、塑胶场地、沙坑等。

　　游憩空间的周围设置了生物滞留池和下沉式花田，大面积的花卉斑块丰富了景观，同时消减了场地周围的径流，使得人们活动的场所不被雨洪所影响。

生物滞留池和花田景观

5　海绵城市与滨水空间

场地内地被植物以耐旱、耐短期水湿、夏季生长旺盛的宿根花卉和观赏草为主，夏季开放的花卉吸引游客驻足观赏，拍照留念

废弃轮胎

废弃沥青

建筑垃圾制作成石笼

从场地中回收的建筑垃圾放置于石笼中，可以达到侧面渗透的作用，通过层台式种植和石笼的缓速，解决坡面径流的流速问题；利用废弃轮胎对场地进行装饰和点缀，为游客提供观赏、休憩的空间；回收的废弃混凝土砌块和砖块也给设计带来了具有活力的形式和丰富的色彩。

迁安是中国的钢铁之城，利用废弃钢材构建特色的、低成本的雕塑和构筑，经过工匠的加工后成为带有时间记忆的艺术小品，运动主题雕塑也为场地带来了活力。

雕塑

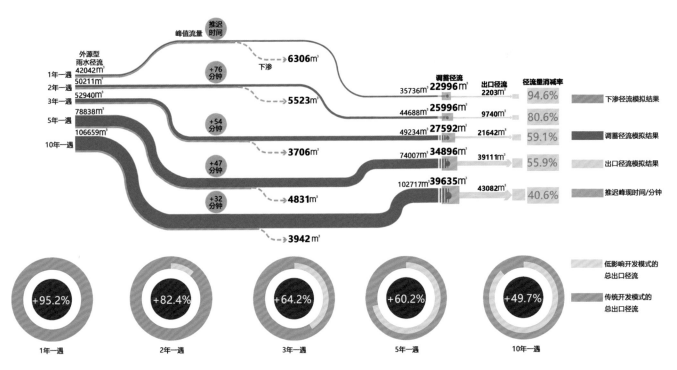

雨水径流处理流程图

雨洪管理能力

设计中利用了SWMM径流模拟软件对方案进行了模拟评估，量化场地对于城市雨洪管理的贡献。

水资源方面：场地可以收集超过2万m^3径流，满足场地内部绿地单次灌溉的用量。

水安全方面：面对外部268.8hm^2的城市径流，场地极大程度地减小了峰值流量，推迟了洪峰时间。

水环境方面：场地对SS、TM、TP、COD等污染物，都具有良好的净化效果。

迁安市滨湖东路东侧绿化带景观工程尝试着用风景园林的方式处理城市雨洪问题，在海绵城市建设和城市绿地基本功能中找到平衡点，在城市绿地科学与艺术的双重价值中找到平衡点。

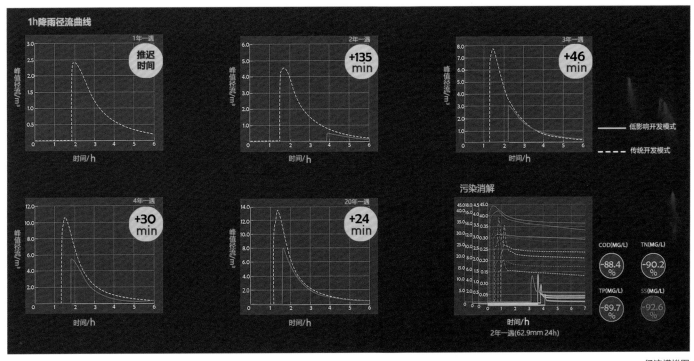

径流模拟图

延展地脉文脉的滨水空间
——河北迁安佛山公园规划设计

A Waterfront Landscape Design that Extends the Natural and Cultural Veins
— Planning and Design of Buddhist Culture Park, Qian'an City, Hebei Province

建成时间：2015 年	Time of completion: 2015
建设地点：河北省迁安市	Construction site: Qian'an City, Hebei Province
项目面积：34.2 hm²	Project area: 34.2 hm²
建设单位：迁安市园林绿化管理局	Construction unit: Qian'an Garden and Greening Administration

获 奖 信 息：2018 年国际风景园林师联合会亚非中东地区文化与传统类杰出奖
Awards: Award International Federation of Landscape Architects Asia-Africa Middle East Region Award (IFLA AAPME) Culture and Traditions Outstanding Award in 2018

主要设计人员： 李 雄　李冠衡　戈晓宇　董 璁　林 洋　肖 遥　林辰松　等
Project team: LI Xiong, LI Guanheng, GE Xiaoyu, DONG Cong, LIN Yang, XIAO Yao, LIN Chensong, et al

　　朝阳初照，青烟混合着雾气在氤氲的石山间袅袅流动；浪石相搏，水声吞吐与晨寺钟声余韵徐歇，古迁安八景"云寺晓钟"恰位于此。巧的是，1500多年来，不管庙宇兴盛还是衰落，伴随着滦河河水的呼吸，这一方土地始终承载着当地居民对佛教文化的尊敬，更凝练着人们对土地、河流、自然的崇拜与热爱。

　　场地位于河北省迁安市城西，钢城大桥西北角，东隔滦河与城市相望，总面积34.2hm²，于2015年建成。迁安的城市区域由于首都钢铁厂搬迁的原因进一步向西扩张，新城规划将跨过滦河沿其西岸发展。从北魏的石造像到面向滦河的庙宇，场地内文化底蕴深厚，1000多年来庙宇从北魏制式变为唐制式，再变为明清制式，甚至短暂消失，从历史到艺术的价值都值得记载与传承。

　　设计师以地域文化的宽度和广度作为弹性框架对地脉进行科学梳理和评估。在延续和拓展场地文脉、地脉的基础上，引入滦河水，梳理和利用现状地形，使场地与滦河共同呼吸，将地域文化与整个自然系统融为一体。除了传承场地的信息与气质，整合并利用风景资源外，还在建设过程中最大程度降低了对地貌、植被、汇水、动物生境的干扰，项目的建成既丰富了城市底蕴，又为城市滨水空间注入新的活力。

Every morning mist with incense curling flow among stones slowly. Waves dash against the river bank, making loud splash which lingers bell bongs fade far away. "Yun Si Xiao Zhong", one of eight scenes in ancient Qian'an has coincidentally been here for over 1500 years. Regardless of the temples' prosperity or decline, along with the breath of water of the Luan River, this land has always been a hot spot of Buddhist culture which is respected by local residents, even has been becoming a symbol of love among people, river and nature.

The project completed in 2015 is located in the west of Qian'an City, Hebei Province, China, at the northwest corner of Gangcheng Bridge, facing the city across the Luan River in the east, with a total area of 34.2 hectares. As the city of Qian'an expands further westward due to the relocation of the Capital Steel Works, the new town is planned to cross the Luan River and develop along its west bank. From the stone statues of the Northern Wei Dynasty to the temples facing the Luan River, the site has a profound cultural heritage. Over the past one thousand years, they have changed from the Northern Wei to the Tang, and then to the Ming and Qing Dynasy, and even disappeared briefly. The historical and artistic values are worth recording and inheriting.

Designers take the width and breadth of regional culture as an elastic framework to conduct scientific sorting and evaluation of geographical veins. Based on the continuation and expansion of the site's culture and geography, the Luan River is introduced to sort out and make use of the current terrain, so that they breathe together, and finally form the integration of regional culture and the whole natural system. In addition to inheriting the information and temperament of the site, integrating and utilizing the landscape resources, the project also minimizes the disturbance to the landform, vegetation, catchment and animal habitat during the construction process. The project not only enriches the urban heritage, but also infuses new vitality into the urban waterfront space.

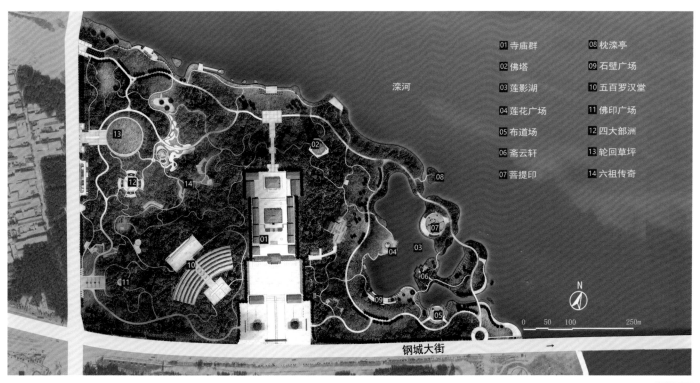

01 寺庙群	08 枕滦亭
02 佛塔	09 石壁广场
03 莲影湖	10 五百罗汉堂
04 莲花广场	11 佛印广场
05 布道场	12 四大部洲
06 斋云轩	13 轮回草坪
07 菩提印	14 六祖传奇

场地平面

实景鸟瞰

5 海绵城市与滨水空间

策略一：理水策略

由于场地为滦河堤外，中间又矗立山体，所以山体西侧、东侧低地一进入雨季便经常内涝，严重影响场地的使用和植被生长。通过坡度分析得出汇水方向、竖向分析得出汇水面积和位置，并结合滦河水位变化，详细计算得到水线范围和水位变化范围，为打开堤坝、融合水系提供科学依据和数据支持。选择高程最低、最易形成水口的东部三段堤坝予以拆除，通过设计干预完成场地内湖与滦河水系的融合，以弹性手段调蓄湖水，形成可控制的水位变化。实现内外水体的自我清洁，促进水中动植物和微生物带动水生态系统的循环，形成有机的生态联系。

堤坝的局部拆除形成水中三洲，契合中国传统自然山水园"一池三山"的格局。内水外水的交融将场地有限的湖面空间扩展至无限广阔。此格局实现了水的弹性供给，解决了水源的可持续问题；形成天然缓冲带，以吸纳雨水的方式减弱极端降雨的危害；设计中以低干扰的方式加速了滦河河道自然修复进程，促进了原有植被系统的演替，形成了新的山水融合的自然结构。此格局实现了水的弹性供给，解决了水源的可持续问题；形成了天然缓冲带，以吸纳雨水的方式减弱极端降雨的危害；设计中以低干扰的方式加速了滦河河道自然修复进程，促进了原有植被系统的演替，形成了新的山水融合的自然结构。

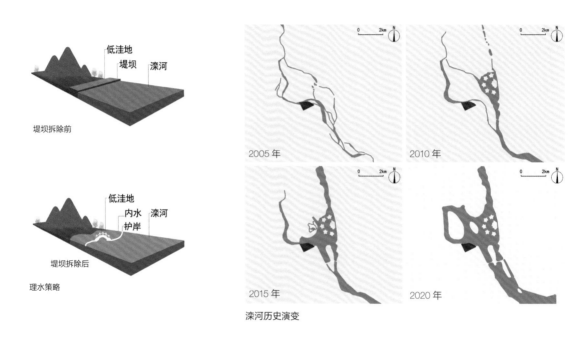

理水策略

滦河历史演变

滦河石桥

策略二：延续地域文化

提取与场地最相宜的当地文化特色，以符号化的设计语言展现在场地的空间格局以及各个节点当中，如场地中部寺院建筑群遵循明清禅宗"伽蓝七堂"制度，沿地形形成由南向北的寺院轴线，为礼佛民众营造充满仪式感的文化氛围。通过对场地及场地构筑的设计，立体地展示地方特色和精神。在场地游赏系统中，塑造一系列的文化活动空间，如莲花广场、菩提印、佛印广场等。表达经典的目的在于使景观表达摆脱历史局限性，典籍中的记述既成为框架，又不指代特定历史节点，使得景观能够更好地应对文化快速发展带来的冲击，利于文化景观的可持续发展。

明清时期的迦蓝七堂格局

兴衰变迁

枕滦亭

寺庙建筑

建成实景

策略三：尊重场地现状

在充分尊重现状地形地貌的基础上，对场地西侧存在的沟壑和自然排水通道加以保留和梳理，规划地表径流和汇水区，形成雨洪管理系统，滋养植被的同时，可变景观的出现增添了场地的生机和意趣。应用滨水植物、乡土植物和中国传统文化植物。以落叶阔叶林为基底，采用复层混交的种植形式，并加入吸引鸟类的植物，增加生物多样性。青瓦铺地、夯土墙、自然石墙、茅草屋顶、木结构建筑等设计细节运用朴素而传统的材料，致敬场地精神。

采用乡土材料　　　植物选择

保留场地奇石

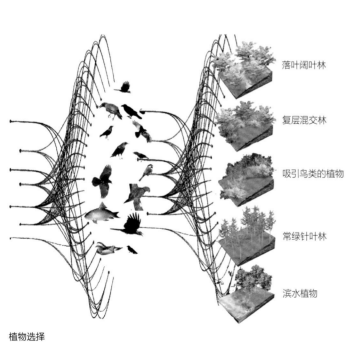

植物选择

保留树石沟

雨水花园体系

策略四：满足活动人群的需求

在场地游赏系统中，设置莲花广场、菩提印、佛印广场、缀花草坪、儿童乐园，滨水景观风光带等一系列活动空间，丰富的活动场地满足了市民各式各样的游憩需求。结合场地特征，对道路游线进行细化和丰富，串联水系，连接文脉，打造多维度的游憩体验，使游客与场地产生更多的情感联系和共鸣，拓展了深层次了解地域文化的空间，给市民生活带来活力与幸福感。

对游赏人群的活动和对场地的态度进行调研分析，发现在建设后各种群体对于场地的评价都有很大提升。场地满足了不同群体的期待，受到了普遍的欢迎，成为居民赏景、休闲、健身的好去处。景观的韧性和活力也随着时间的流逝愈发完善。

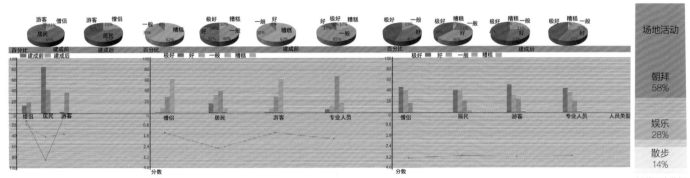

人群活动分析

儿童乐园

滨河带风光

滨河广场

滨河步道

传承晋商文化的绿色枢纽
——山西晋中晋商文化公园规划设计

A Green Hub for Inheriting the Culture of Shanxi Merchants
— Planning and Design of Jin Commercial Cultural Park, Jinzhong City, Shanxi Province

建成时间：2011年	Time of completion: 2011
建设地点：山西省晋中市	Construction site: Jinzhong City, Shanxi Province
项目面积：25 hm²	Project area: 25 hm²
建设单位：晋中市园林局	Construction unit: Jinzhong City Landscape Bureau

获 奖 信 息：2018年国际风景园林师联合会亚非中东地区经济价值类荣誉奖
2017年中国勘察设计协会计成奖二等奖
Awards: Award International Federation of Landscape Architects Asia-Africa Middle East Region Award (IFLA AAPME) Economic Value Honorable Mention in 2018
Second prize of China Engineering & Consulting Association Jicheng Award in 2017

主要设计人员：李 雄　董 璁　瞿 志　李运远　蔡凌豪　姚 朋　戈晓宇　郑晓东　林 洋　郑 曦　李冠衡　张云路　林辰松　肖 遥　李方正　等
Project team: LI Xiong, DONG Cong, ZHAI Zhi, LI Yunyuan, CAI Linghao, YAO Peng, GE Xiaoyu, ZHENG Xiaodong, LIN Yang, ZHENG Xi, LI Guanheng, ZHANG Yunlu, LIN Chensong, XIAO Yao, LI Fangzheng, et al

　　晋商文化公园位于山西省晋中市，公园位于新城区的南部，老城区的北部，是新老城区的过渡区域，也是重要纽带。晋商文化公园是一个长约1.5km，宽150~250m，占地约25hm²的绿带，由4块场地组成，被3条市政道路相隔。该区域过去主要是农田，地形平坦，区域内遍布农民倾倒的垃圾，导致了场地景观类型的单一以及生态环境的破坏。

　　项目的设计策略体现在经济、生态和文化方面。将原来垃圾填埋场为主的环境规划为一个著名的文化中心和生机勃勃的新兴绿地。项目将旧城区的经济和社会资源吸引到新城区，随之变化的是居住环境改善，产业结构调整，居民收入增加，土地价值提升，最终形成良性循环，吸引更多的人迁入新城区，从而全面提升区域经济发展。

　　公园建成已超过5年，西侧的两块场地与城市规划馆、图书馆和博物馆等公共建筑相邻，公园的其余部分与晋中市核心的居住区紧密相连，有小区和商业街，并配备了相应的公共设施、商业设施和绿地。

　　Jin Commercial Cultural Park is located in Jinzhong city, It sits at the north of old town, the south of the new town, which is the transition area of new and old town and an important coupling of new zones as well. Jin commercial cultural park is a belt of green space, which is about 1.5 km long and 150 m to 250 m wide, covering an area of approximately 25 hm². The park is composed of four pieces of land which is separated by three municipal roads. The past situation of this area was mainly covered by farmland of corns and wheat, and the terrain is flat. The local area is suffering from garbage dumping by the surrounding farmers, which contributes the fact of simplified landscape type and poor ecological environment.

　　The design strategies of this project drive from the aspects of economy, ecology and culture. It changed the original environment—landfill in this area into a prominent cultural center and a vibrant emerging green space. With the living environment improvement brought by this project, it draws the economic and social resources of the old town into the new town, adjusts the industrial structure in this area, increases residents' income, and promotes the land value in the region. Eventually, it forms a virtuous cycle and attracts more people to move in, and then comprehensively improves the regional economic development.

　　Jin commercial cultural park has now been completed more than 5 years. The 2 pieces on the west side are blended with public institutions such as the City Planning Hall, library and museums. The rest of the park is linked closely to the most expensive real estate in JinZhong City, which contains residential areas and commercial streets and has equipped with public facilities, commercial and beautiful greenery.

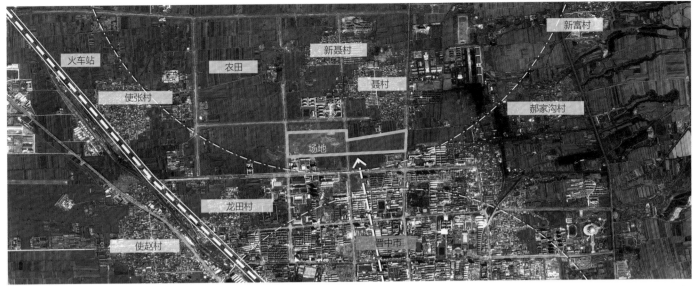

区位分析

晋商公园位于新城区南部，旧城区的北部，是新旧城区的过渡区域，也是重要的纽带。公园距离市中心近3km，北接晋阳街，南临荣军北街，东西两侧均为规划路，作为协调区域发展的枢纽，对城乡一体化发展起到了关键性的作用。

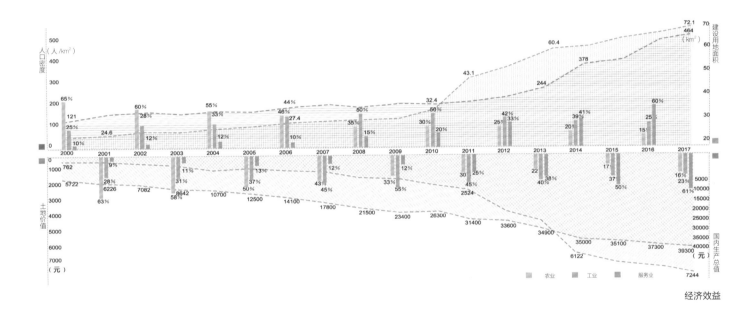

经济效益

文化方面，晋商公园每年至少举行12次文化节，包括1月的社火节，5月的戏剧节等。最大的狂欢节是社火节，有1万人参加；生态方面，随着晋商公园的建设，植物种类增加至200多种。

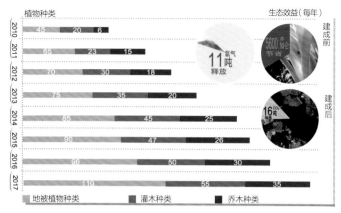

生态效益

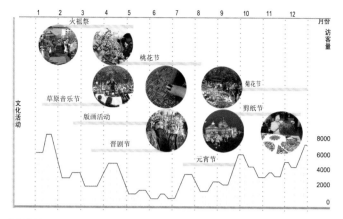

文化效益

01. 庭院印象
02. 岛屿
03. 亭
04. 花园
05. 晋中城乡规划局
06. 晋式亭
07. 停车场
08. 贵宾停车场
09. 餐厅
10. 咖啡及零售
11. 文化展馆
12. 公共厕所
13. 亲水平台
14. 儿童乐园
15. 码头剧场
16. 晋文化景观带
17. 瀑布
18. 广场
19. 户外剧场
20. 停车场
21. 服务中心
22. 停车场
23. 公共厕所
24. 读书亭
25. 垂钓亭
26. 茶室
27. 风筝草坪
28. 停车场
29. 健身场
30. 停车场

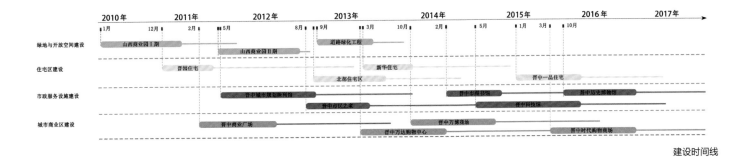

建设时间线

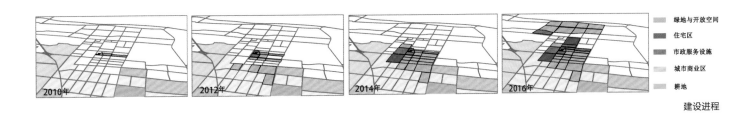

建设进程

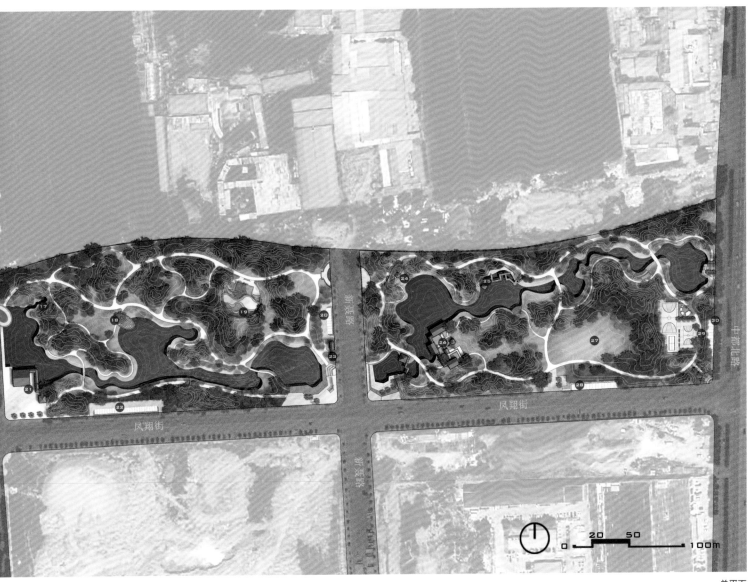

总平面

5 海绵城市与滨水空间

在美丽中国的大背景下，晋商文化公园作为新旧城区协调发展的枢纽，将旧城区的经济和社会资源吸引到新城区，改善城市生态环境，促进区域交流发展，传承城市历史文脉，在经济、文化、生态等方面发挥了重要的作用，是宣传城市美丽形象的一个窗口

公园将历史文化中的元素进行场地适应性转译，并融入景观设计中，使用晋商四合院通常使用的灯笼符号作为广场照明；使用传统建筑的砖雕作为广场铺装，突出了区域特色。

整个公园都连接了雨水系统，以确保雨水的收集。集雨水收集、水景建设和绿地灌溉于一体，公园四周沿街道不设围栏，与商业区相连的区域被设计为连续的开放空间以及公园外公共的休息空间

传承中国传统的园林理法将传统建筑融入自然

晋商公园建设前后，随之变化的是生态系统稳定，居住环境改善，产业结构得到调整，居民收入增加，土地价值得到提升，吸引更多的人迁入新城区，从而全面改善区域经济发展，最终形成经济、文化、生态多方面的良性循环

滨水景观通过设置平台栈道，满足游人亲水的需求，创造有趣的滨水空间和动物栖息地形成了一个稳定的生态系统

5 海绵城市与滨水空间

炎帝文化复兴的生态文化廊道
——山西高平炎帝文化公园规划设计

Ecological Cultural Corridor of Yandi Cultural Renaissance
— Planning and Design of Yandi Cultural Park, Gaoping City, Shanxi Province

建成时间：2016 年	Time of completion: 2016
建设地点：山西省高平市	Construction site: Gaoping City, Shanxi Province
项目面积：15 hm²	Project area: 15 hm²
建设单位：高平市南城街街道办事处	Construction unit: Nancheng Street Sub-district Office in Gaoping
获奖信息：2019 年国际风景园林师联合会亚太地区文化及城市景观类荣誉奖	Awards: Award International Federation of Landscape Architects Asia-Pacific Region Award (IFLA AAPME) Cultural and Urban Honorable Mention in 2019
主要设计人员：李雄 姚朋 孙漪南 戈晓宇 林辰松 胡楠 等	Project team: LI Xiong, YAO Peng, SUN Yinan, GE Xiaoyu, LIN Chensong, HU Nan, et al

　　炎帝公园位于山西省高平市南城街，设计用地由两部分地块组成，总面积约 15hm²。高平市以前有 35 座帝王庙，自金代开始横跨 6 个朝代，贯穿 900 多年历史，但是现仅存 3 座，且不在市区。城市河流丹河穿越场地，场地内部有废弃建筑及运动场地等；现有树种主要为杨树、柳树、雪松，树种长势良好，种类较为单一。

　　在炎帝文化的大背景下，炎帝公园寻求自身特色与其他纪念性场所的互补，通过炎帝故事的诉说、炎帝文化建筑与场所的营造，打造主题鲜明的景观。公园依托丹河形成连续的水域生态廊道，为人们的绿色出行、动植物多样性、区域环境改善等提供平台，发挥城市绿色基础设施的积极作用。在建设美丽中国的大背景下，公园通过绿地景观和文化建设引领区域发展，促进和带动周边地块和产业结构的优化，成为新时期生态文明和美丽高平建设的绿色引擎器。

　　炎帝公园以开放空间的形式位于高坪市中心的丹河河畔。公园与炎帝陵墓、阳头山、炎帝文化遗迹相连，在高平市区形成一个文化网络。公园承载着不同的文化和日常活动，激发城市的活力，并重塑炎帝的文化信念，它的建设对炎帝文化的复兴和城市活力的激发具有重要意义。

　　Yandi Park is located in Nancheng Street, Gaoping City, Shanxi Province. The design area is composed of two parts, with a total design area of approximately 15 hm². In Gaoping city, there used to be 35 Emperor Yan temples, from the Jin Dynasty across 6 dynasties with a history of 900 years. But now there are only 3 exist and none of them in the downtown area. The city river Dan river passes through the site, and there are abandoned buildings and sports fields inside the site. The existing tree species are mainly poplar, willow and cedar, with good growth and relatively single species.

　　Under the background of Yandi culture, we seek the complement of the characteristics of the venue with other commemorative venues. Through the narration of Yandi's story and the construction of Yandi's cultural buildings and places, we create a park landscape with a distinctive theme. Relying on the Dan River to form a continuous water ecological corridor, the park provides a platform for people's green travel, the construction of animal and plant diversity, and the improvement of regional environment, so as to play positive effects of urban green infrastructure. Under the background of building a beautiful China, the park lead the development of the region through green landscape and cultural construction, promote and drive the optimization of surrounding land parcels and industrial structure, and become a green engine for ecological civilization and beautiful Gaoping construction in the new era.

　　Emperor Yan Park is connected to the Emperor yan's Mausoleum, Yangtou Mountain, Yandi Culture Relics, etc. forming a cultural network in the downtown area of Gaoping. The completion of the park carries different cultural and daily activities, which stimulates the city vitality and reshapes the culture beliefs of Emperor Yan. Its construction is of great significance to the revival of Yandi culture and the stimulation of urban vitality.

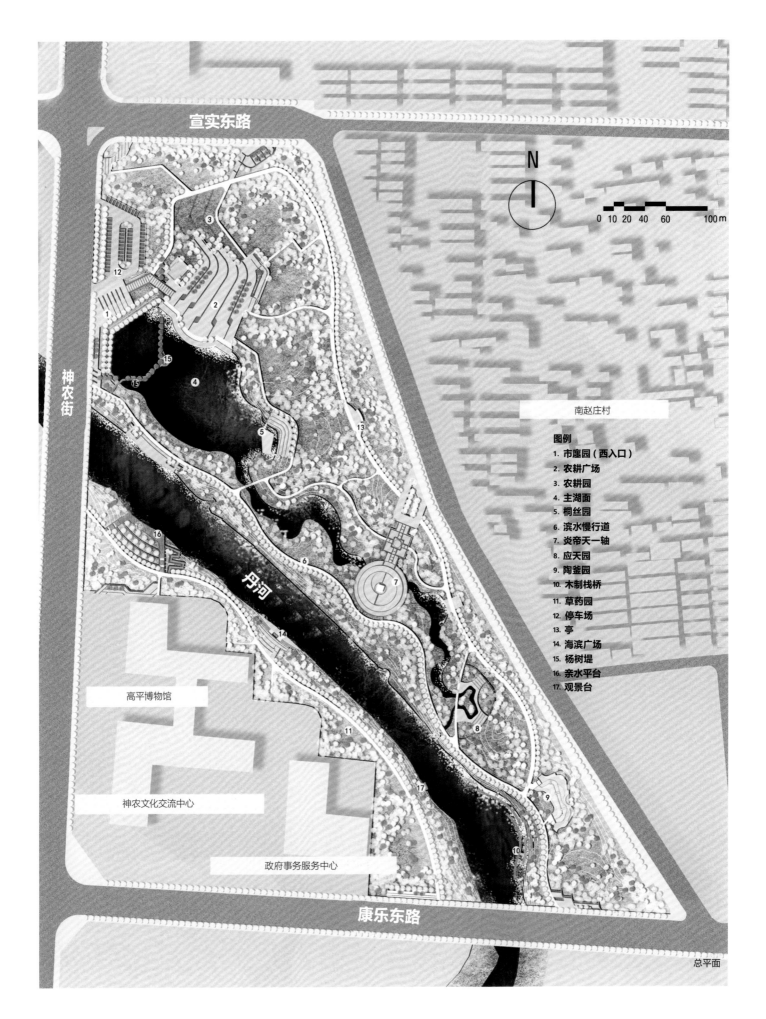

总平面

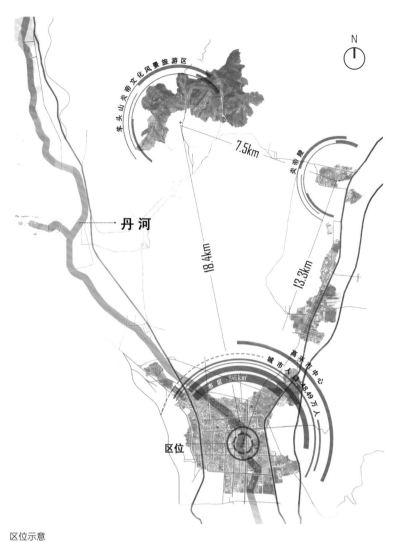

区位示意

炎帝是中华民族的始祖，又称神农，是中国三皇五帝的领袖。他带领其氏族耕种，取代了狩猎和捕鱼，并驯化了许多植物，发明了许多农具，还教人们如何贮藏食物。他创造了中国的农业文化，积累了很多草药知识，其精神持续了数千年，延绵至今。

高平市位于中国中部平原的山西省南部，是炎帝曾居住和耕种并最终被埋葬的地方，承载炎帝文化的高平市是中国最重要的纪念炎帝文化的城市。

丹河——高平市的母亲河，贯穿城市，并连接高平的诸多炎帝文化遗址；炎帝公园位于市中心，丹河河畔。

季节性湿地

公园内保留了 28 棵现状大杨树，使它成为分割公园核心水面的长堤，丰富了水景层次，并在水岸边营建亲水栈道，让整个滨水地带充满了活力。通过种植观赏草，并搭配乔灌木的植被，丰富植物结构层次，突出植物景观的原生态性、自然性、低维护性，营造多样性的植物风貌。

湖心杨柳栈道

在雨季，公园内的季节性湿地起到储蓄雨水、净化水质以及涵养地下水的作用，通过种植芦苇等水生、湿生植物，形成四季变化的生态景观，同时生长良好的芦苇吸引了鱼类、水鸟栖息，不仅营造了良好的生物栖息地，提高生物多样性，还对恢复场地的生态系统起到了重要的作用。季节性湿地还赋予了场地科普教育的意义，通过设置科普休闲设施，湿地成为孩童们嬉戏的天堂。

季节性湿地

湿生景观

建筑与植物相映成趣，共绘成一幅城市自然图卷；城市与公园景观融为一体，为文化活动和娱乐提供了场所；驳岸软硬结合，既满足了游客的停歇，又营造了多层级的景深。

桐丝园

桐丝园以水为弦,运用水剧场表达炎帝造琴的音乐主题。通过竖向设计和不同层次的种植,结合水剧场功能,形成富于变化的停坐空间,与湖区的水景相映成趣。

季节性湿地

卫星城区催化剂
——山东济阳安澜湖水公园
Catalyst of Satellite District
— Anlan Lake Park, Jiyang City, Shandong Province

建成/完成时间：2017年	Time of completion: 2017
建设地点：山东省济南市济阳区	Construction site: Jiyang District, Jinan City, Shandong Province
项目面积：7.2 hm²	Project area: 7.2 hm²
建设单位：山东建大景园科技有限公司园林规划设计中心	Construction unit: Garden Planning and Design Center, Shandong Jianda Jingyuan Technology Co., Ltd.
主要设计人员：李 雄 郑 曦 戈晓宇 段 威 胡 楠 等	Project team: LI Xiong, ZHENG Xi, GE Xiaoyu, DUAN Wei, HU Nan, et al

黄河安澜，是中华儿女的千年期盼。在这片黄河两岸的历史、文化和自然力量的激荡、融合与组织中，孕育出具有独特属性的城市群。济阳，作为黄河下游中心城市济南北跨发展战略的先行示范区，与济南城中心隔黄河而望，承担着疏解济南城市压力、协调城市高质量发展的重要职能。

项目面积共72000m²，场地位于济阳城市中轴线与蓝绿生态轴线的交汇处，黄河引水渠穿园而过。设计目标是建立彰显黄河文化的城市景观，通过生态技术缓解城市排洪压力，最终带动城市边缘区发展。项目作为济阳新城与黄河区域性联系的中心动脉，周边汇集城市文体中心和市政中心等公共服务设施。而作为新城绿地系统建设的关键，设计团队思考着如何通过风景园林的方法，恢复和重新诠释黄河带来的文化与生态馈赠。

The peaceful flow of the Yellow River is the thousand-year expectation of the Chinese people. In the agitation, integration and organization of historical, cultural and natural forces on both sides of the Yellow River, a city cluster with unique attributes has been bred. Jiyang, as the pilot demonstration area of the north trans-development strategy of jinan, the central city of the lower Yellow River, and the city center of Jinan across the Yellow River, undertakes the important functions of relieving the pressure of Jinan city and coordinating the high-quality development of the city.

The project covers a total area of 72000 square meters. The site is located at the intersection of jiyang city's central axis and blue-green ecological axis. The Yellow River diversion canal runs through the park. The design goal is to create an urban landscape that reflects the Culture of the Yellow River, alleviate the pressure of urban flood discharge through ecological technologies, and ultimately drive the development of urban fringe areas. As the central artery of regional connection between Jiyang New City and Yellow River, the project gathers public service facilities such as urban sports center and municipal center around it. As the key to the construction of the green space system in the new city, the design team thought about how to restore and reinterpret the cultural and ecological gifts brought by the Yellow River through the method of landscape architecture.

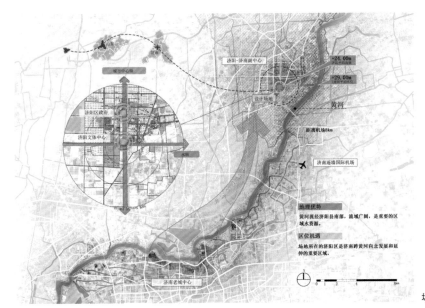

本项目坐落于济阳城市中轴线和水轴线的交叉点,场地内北侧有黄河引水渠穿过。三面环路,位置优越,北接城市轴线地标性建筑文体中心。

场地区位

地理优势:黄河边的水资源城市

由于黄河河床与城市平原有着天然的高差,造成区域地下水位极高,为城市提供了大量的水资源,并逐渐发展出引黄灌溉等传统的人居智慧,同时造就了济阳多处与黄河水文化相关的历史景点。

区位机遇:高密度城市中心北跨黄河的先行区

公园的设计愿景围绕着人们的生活而展开,尊重当地自然环境与文化属性。基于生态空间的律动发展,以及与城市和黄河文化的深度融合,安澜湖公园在城市发展与生态适应之间构建互哺互融的完美共生关系。通过雨洪管理、边界赋能、文化感知、生境共建四个策略,将安澜湖公园塑造为城市蓝绿基础设施的关键核心,使其成为促进地区高质量发展的催化剂,催化济阳区的社会生态发展,为居民提供一个幸福、美丽和高品质的全新城市生活。

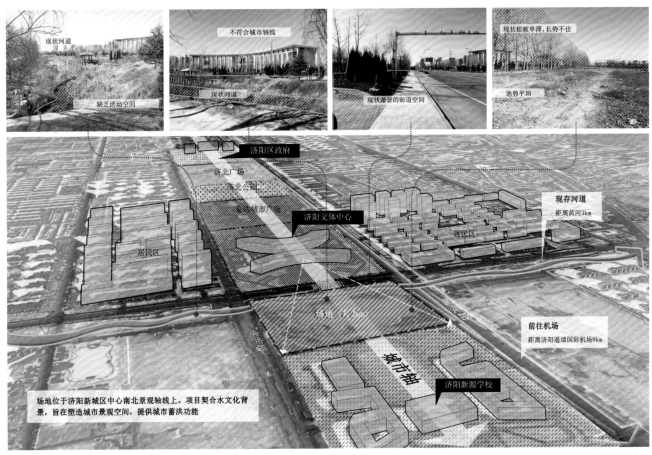

周边环境分析

总平面

项目效益

建成以来,安澜湖公园已经成为济阳区的公共活动中心,为城区增添了活力,提升了人们的生活品质。公园的雨洪管理功能对城市可持续发展有着生态意义。更重要的是,安澜湖公园以水为媒,将场地形式、功能以及独特的文化深深地交织在一起,让城市居民从使用水到观赏水、体验水到感悟水,在公园中能真正体会到黄河对城市发展带来的深刻影响。如今的安澜湖公园已经成为迅速扩张的城区中最受欢迎的公共空间,同时也是济阳区身份特征与文化形象的核心所在。未来,安澜湖公园将成为黄河流域城市高质量发展的全新典范。

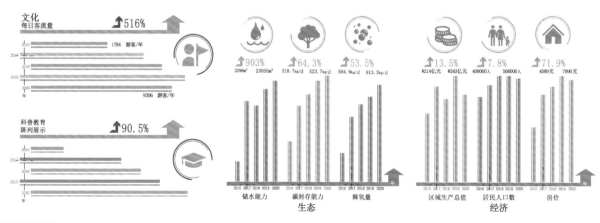

项目效益分析

策略 1：水系统与雨洪管理

场地紧邻黄河，地下水位低于地表不到 1 米，功能上要求满足一定的滞洪需求。

因此设计团队汲取黄河传统人居智慧，凭借场地原有的地势地貌与人文特征，构建充满韧性且富有活力的水系统。园内的水系连接现有河渠，通过一系列集雨设施收集雨水，并以多种净水湿地过滤净化，改善水质，最后汇入中心湖体，进行储存与利用，形成可持续的全园水循环系统。公园的雨洪管理系统也积极融入城市水网体系，成为城市蓄洪滞洪弹性管理的重要途径。

策略 2：城市与公园的双向赋能

场地从一片荒地转换为城市景观轴线上功能完善的公共空间，打破传统公园与城市的界限，创造出交织的尺度和肌理，需要与周边城市公共建筑和街道空间建立动态联系。北侧滨水广场，连接城市文化设施；东侧道路是济南国际机场路直达济阳市中心的必经之路，塑造街景与休憩空间；南侧湖滨森林，为学校营造静谧氛围；西侧八景园与儿童活动场，服务周边社区居民。公园内设置多处观景构筑，中轴南端打造标志性的观景台，作为公园内的制高点，让视线远眺湖光风景，开辟城市全新视野。

策略 3：水文化的场景化感知

依托丰富的水系统骨架，公园营造出滨水场所和文化景观的多样肌理。设计充分利用场地水资源丰富的天然优势，营造湖泊、喷泉、瀑布、浅滩、湿地等动态与静态水景交织的充满节奏的场景式体验。通过净水瀑布连接湖面与水渠，平远悠扬的湖面与北侧文体中心交相辉映。打开北侧公园城市边界，设计城市滨水广场沿路欣赏水体景色。在湖面南侧营造出水岸游憩与林中漫步的环境，设置临水茶室、环形栈桥等，形成湖滨休闲活动中心。公园的水净化技术也作为景观的可见部分被展示，如石笼、滩石以及自然驳岸等净化设施，创造出自然科普的室外课堂。

策略 4：生境共建共享

在对济阳的动植物群落特征与习性等研究的基础上，方案串联了丰富的自然生态系统要素，打造浅滩湿地、水岸疏林、森林草坡等生境类型，建立多层次的生态网络。团队选用柳树、千屈菜等具有净化能力的耐水湿植物，提升浅滩湿地的净化能力，并结合菖蒲、睡莲、慈姑等湿生植物在环形栈桥中进行科普展示，打造户外自然教育和交流空间。

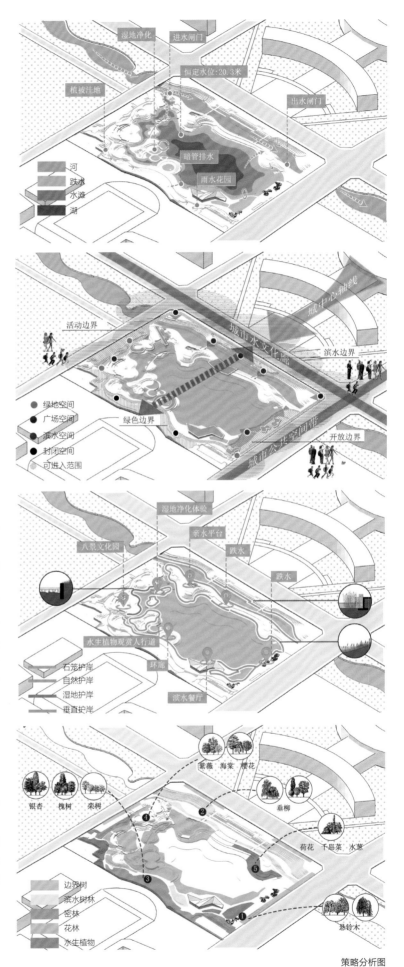

策略分析图

城市生活

滨水休闲

科普教育

活动类型

场地活动设置展示

各年龄层的丰富活动、绵延交织的文化、生态网络吸引着周边各年龄层的居民，成为探索自然奥秘的绿色中心。

北部河渠建设前后

北部河渠恢复了自然驳岸，种植了更多灌草植物增加观赏性和净水能力，从建设前单一的景观形式，成为文体中心对面美丽的城市河岸景观。

改造前

河道生态化改造后

步道体系

健身步道

公园设置了一条600m长的健身跑道环绕公园,河渠和湖体之间的景观堤是连接公园东西的重要通道。

水净化景观展示

展示石笼、滩石以及自然驳岸等多种形式的净化措施以达到科普教育的目的,公园的水净化技术作为景观的可见部分被展示。

水净化

5 海绵城市与滨水空间

降雨通过渗透的方式进入跌水台第一层，依次跌入下层；跌水台共分3层，每一层栽种不同的水生植物，能够净化水质，起到分级净化的作用，跌水的落差也能给水中提供一定的氧气。

水净化景观展示——跌水

部分驳岸采用石笼的形式结合种植，兼具生态和景观功能。雨水径流流入水体前，先流经位于驳岸的石笼，石笼填充物和植物共同发挥了截留径流中污染物的作用。同时石笼种植的植物，通过光合作用释放到水体中的氧气也有益于水体净化。

水净化景观展示——石笼

采用湿地植物的优化组合，增强湿地对暴雨径流以及相关污染物的截留与过滤能力，提高了对水中氮磷的净化效果，使整个生态系统高效运行，最终形成稳定可持续利用的系统，提高了系统净化河水的能力。

水净化景观展示——浅滩湿地

"树环"可使游客在顶部眺望风景,并在底部近距离的观赏湖面

依托丰富的水景,公园内有多处景观节点供游客体验水文化的印记,感受不同尺度的水景空间。湿生植物在环形栈桥中进行科普展示。

湿生植物科普展示

栈桥上层空间

栈桥内部空间

5 海绵城市与滨水空间 183

瞭望城市轴线

公园东部是开放的街景空间,作为南北绿轴上步行体系的一部分,这里视线开阔,因此,公园东部设计为开放街景空间,视线通透,可以从道路上欣赏湖体风光。

公园内设计的安澜湖,以开阔的湖面风光成为城市景观轴线上的重要节点。

公园设计了多处构筑作为核心节点供人们体验水文化的印记;植物作为生态措施之一,多种类的湿生植物和耐湿品种结合水系统和驳岸进行种植。

城市街景空间

河岸植物生境

实景鸟瞰

湖滨景观建筑

6
城市更新与绿色开放空间
Urban Renewal and
Green Open Space

改革开放四十多年来，我国的城市建设经历了快速扩张时期，城市空间日趋饱和，土地资源愈发紧张，城市更新逐渐成为我国城市空间发展的新方式。风景园林为了顺应新时代改善人居环境和重塑城市活力的发展需求，积极探索城市更新的范式，科学提高城市更新的绩效。这既是新时代风景园林必须履行的社会责任，也是专业发展必须面对的更高的新要求。

城市更新是绿色空间的更新

城市发展不是孤立的，城市与周边自然共同构建了一个和谐共生的人居系统。城市更新工作需要重新建立城市与自然之间的有机融合关系。在更新中，风景园林要进一步强化"生态优先、绿色发展"的基本理念，以城市生态空间的保护修复作为城市空间更新的基础，促进城市生态系统与人居系统耦合协调。通过"保绿"推进城市绿色资源的保护和生态系统的修复，重新塑造城市与自然的共融关系；通过"引绿"实现城市外围绿色空间的渗透和链接，重新建立城市与生态的物质循环机制；通过"还绿"高效利用城市废弃空间和腾退闲置土地，重新塑造城市的绿色开放空间体系。

城市更新是环境品质的更新

实施城市更新行动的宗旨，是推动解决城市发展中的突出问题和短板，提升人民群众的获得感、幸福感、安全感。这意味着以基础设施更新优化为中心，以老旧硬件设施升级为重点的发展思路，变成了以人民为中心，以生活质量为导向的城市环境品质综合提升模式。风景园林可以让城市释放、营造和重组更多绿色宜居、健康人本的绿色开放空间，彰显城市更新对美好生活和民生福祉的促进作用。

城市更新是生态韧性的更新

雨洪压力、空气污染、地质灾害、地震火灾等灾害在威胁着城市安全。城市更新的目的不仅是让城市更美、更整洁，重建和恢复一个生态韧性、弹性调节的城市也是当前城市更新工作的一项重要目标。从"景观的高颜值"到"生态的高品质"，通过在城市更新中塑造多尺度、多类型的韧性绿地空间，重建城市生态系统，优化生态绩效，促进破碎地区的生态修复，完善城市防灾避险功能，不断增强城市在承受各种扰动时能够化解和抵御外界冲击的能力。

城市更新是文化活力的更新

文化是城市的灵魂，城市在发展过程中形成的特有历史文脉和文化印迹，是城市气质的重要体现。城市更新可以让城市物质空间以"旧"变"新"，可以让"脏乱差"变成"洁净美"。但在空间物质"新陈代谢"的过程中，不能忘记城市的"文化之根"。延续好城市文脉传统，激活出城市文化活力，实现城市更新的真正价值，这是风景园林的重要使命。一方面，在城市更新中应着重保护和保存历史建筑和历史街区，修复和营造唤起"乡愁记忆"的传统景观风貌和空间场景。另一方面，应致力于"文化遗产活化"，营造识别度高、吸引力强、文化内涵深厚、个性魅力独特的人文交流共享空间。

　　风景园林学科及相关学科在实施城市更新行动中责无旁贷，我们要在全面建设社会主义现代化国家新征程中，坚定不移实施城市更新行动，推动城市高质量发展，努力把城市建设成为人与人、人与自然和谐共处的美丽家园。

　　本章中的各个项目都是在城市更新中保留或恢复的绿色开放空间。

　　晋中139铁路公园是在晋中市绿地系统规划中，规划设计团队敏锐识别的城市废弃铁路沿线的潜力空间，并通过调整城市总体规划，将这里塑造成了老城区中宝贵的绿色空间。

　　晋中市社火公园的场地原状是一处被城市包裹的苗圃，在社火文化获批国家级非物质文化遗产后，这里被定位成了展示非物质文化遗产的绿色开放空间，也是晋中市老城区的一处宝贵的新增公园绿地。

　　西宁五一路口位于非常繁忙紧凑的城市核心区，原有建筑的拆除为市民开辟出了难得的绿色空间，这里承接了立体的人行交通功能，也承载了老百姓的户外绿色休闲功能。

　　通海桥铁路公园处于城市道路和高架桥的包裹之中，又被废弃的城市铁路穿越，是一处依托城市灰色基础设施打造的绿色空间，也是西宁第一个以铁路文化为特色的公园绿地，人们通过对于文化的认知，感受到了城市发展和产业转型带来的巨大变迁，也能够在生硬的城市空间中感受到宝贵的自然魅力。

Over the past 40 years of reform and opening-up, China's urban construction has experienced a period of rapid expansion, with urban space becoming increasingly saturated and land resources becoming more and more tense, and hence urban renewal has gradually become a new way of urban space development in China. In response to the development needs of improving the living environment and reshaping urban vitality in the new era, landscape architecture accurately grasps the connotation of urban renewal, actively explores the paradigm of urban renewal, scientifically improves the performance of urban renewal to meet people's needs for a better life. This is not only the social responsibility of landscape architecture in the new era, but also the new and higher requirements for professional development.

Urban renewal is the renewal of green space

Urban development is not isolated; on the contrary, a city and its surrounding natural environments should be integrated into a harmonious and symbiotic human settlement system. Urban renewal needs to re-establish the systematic integration of city and nature. In the process of renewal, landscape architecture should further strengthen the basic idea of "ecological priority and green development", take the protection and restoration of urban ecological space as the basis of urban spatial renewal, and promote the coupling and coordination between urban ecosystem and human settlement system. Measure should be implemented to promote the protection of urban green resources and ecosystems restoration, and re-shape the relationship between city and nature through "preserving green"; to realize the penetration and linkage of green space in the periphery of the city, and re-establish the material circulation mechanism of city and ecology through "introducing green"; and to make efficient use of abandoned urban space and vacated idle land, and reshape the green open space system of the city through "returning green".

Urban renewal is the renewal of environmental quality

The purpose of implementing urban renewal is to promote the solutions to highlighted problems and shortcomings in urban development and to enhance the people's sense of happiness and security. This means that the development idea of infrastructure renewal and optimization, focusing on the upgrading of old hardware facilities has shifted into a comprehensive people-centered improvement mode of urban environment with emphasis on life quality. Landscape architecture can facilitate a city to release, build and reorganize more open spaces that are green, livable, healthy and human-oriented, showing how urban renewal can promote good life and well-being of people.

Urban renewal is the renewal of ecological resilience

Intensive rain and flood pressure, air pollution, geological hazards, earthquake, and fire are threatening urban safety. The purpose of urban renewal is not only to make the city more beautiful and tidier, but also to rebuild and restore an ecologically resilient and flexible city, which is also an important goal of urban renewal. Landscape architecture helps to rebuild urban ecosystem, optimize ecological performance, promote ecological restoration in broken areas, perfect the function of urban disaster prevention and risk avoidance, and constantly enhance the ability of cities to defuse and resist external shocks in face of various disturbances by shaping multi-scale and multi-type resilient green space from "gorgeous appearance of landscape" to "high quality of ecology".

Urban renewal is the renewal of cultural vitality

Culture is the soul of a city, and the unique historical context and cultural imprint formed by the city in the process of development is an important embodiment of the urban temperament. Urban renewal can change the urban physical space from "old" to "new", and from "dirty mess" to "clean beauty". But in the process of space material "metabolism", we cannot forget the city's "cultural root". It is an important mission of landscape architecture to continue the tradition of urban context, activate the vitality of urban culture and realize the real value of urban renewal. On one hand, in urban renewal, we should focus on protecting and preserving historical buildings and blocks, repairing and restoring traditional landscape features and spatial scenes that arouse nostalgia. On the other hand, we should be committed to the "rejuvenation of cultural heritage" and create a sharing space for humanistic exchanges with high recognition, strong attraction, profound cultural connotation, and unique personality charm.

Landscape architecture and related disciplines are duty-bound in the implementation of urban renewal. In the new journey of building a socialist modern country in an all-round way, we must unswervingly push forward urban renewal, promote high-quality urban development, and strive to build the city into a beautiful home where man and nature coexist harmoniously.

Each project in this chapter is a green open space reserved or restored in urban renewal.

During the planning of the green space system in the Jinzhong City, the planning and design team acutely identified the potential of the space along the abandoned railway line in the city and made it a valuable green space in the old urban area — the Jinzhong 139 Railway Park, by adjusting the urban master plan.

The site of the Jinzhong Spring Festival Parade Culture Park is a nursery enclosed in the city. After the Spring Festival Parade Culture was entitled as National Intangible Cultural Heritage, it was positioned as a green open space to display intangible cultural heritage. It is also a valuable new park green space in the old urban area of Jinzhong.

The Xining Wuyi intersection is located in the very busy and compact urban core area. The demolition of the original building has opened a rare green space for the citizens. It has undertaken the tridimensional pedestrian traffic function and realized the outdoor green leisure function for the common people.

The Tonghaiqiao Railway Park is surrounded by urban roads and viaducts and traversed by abandoned urban railways. It is a green space built on urban gray infrastructure and the first park green space in Xining characterized by railway culture. Through the cognition of culture, people feel the great changes brought by urban development and industrial transformation and can feel the precious natural charm in the stiff urban space.

从棕地到绿色基础设施
——工业城市转型视角下的山西晋中铁路公园更新设计

From Brownfield to Green Infrastructure
— The Renewal Design of Railway Park from the Perspective of Industrial City Transformation, Jinzhong City, Shanxi Province

建成时间：2021年
建设地点：山西省晋中市
项目面积：7.5 hm²
建设单位：晋中市园林局

Time of completion: 2021
Construction site: Jinzhong City, Shanxi Province
Project area: 7.5 hm²
Construction unit: Jinzhong Garden Bureau

主要设计人员：李 雄 张云路 肖 遥 段 威 等
Project team: LI Xiong, ZHANG Yunlu, XIAO Yao, DUAN Wei, et al

  作为中国重要的传统工业基地，山西省晋中市正经历着城市的重要变革。由于多年的工业开发，晋中生态污染，城市空间破碎，社区失去活力，亟待转型。本项目以晋中城市空间为研究对象，希望通过科学的分析和评估，寻找提升城市空间品质的契机。在城市空间大数据统计的辅助下，识别出一段具有综合提升潜力和广泛影响力的废弃工业铁路作为晋中城市更新的媒介。项目从社区活力再生、生态网络构建、工业文化复兴三大策略出发，以废弃铁路绿色更新为主题，重新构建废弃场地与城市的关系，提高居民与绿地的可达性和连接性；塑造多样的动植物生境，修复城市生态系统；并利用遗留铁路设施，延续场地文化记忆，实现"由棕色到绿色基础设施"的规划愿景。

  该项目具有综合的自然及人文效益，并通过多方参与的工作机制，取得了极具影响力的示范成果，为工业城市更新提供了一种可推广、可持续的方法。

  As an important traditional industrial base in China, Jinzhong City in Shanxi Province is undergoing major transformations. Due to years of industrial development, Jinzhong has been left with ecological pollution, fragmented urban space, and communities without vitality. The project takes Jinzhong's urban space as the research object, aims to explore a way to improve the quality of urban space through scientific analysis and evaluation. With the big data statistics of urban space, a section of abandoned industrial railway with comprehensive potential and extensive influence was selected as a medium for Jinzhong's urban renewal. The project starts from three major strategies: the regeneration of community vitality, the construction of ecological network and the revival of industrial culture. The project takes the green renewal of the abandoned railway as theme, reconnects the abandoned area and the city, improves green land's utility and people's connection to it, creates diverse habitats for plants and animals, and advances the restoration of urban ecosystem. The project also takes the railway relics to extend its cultural memory, to realize the vision of "From Brownfield to Green Infrastructure".

  With its comprehensive natural and human benefits, the project has become an influential demonstration through multi-party participation. It provides for the industrial cities a sustainable and promotable way of transformation.

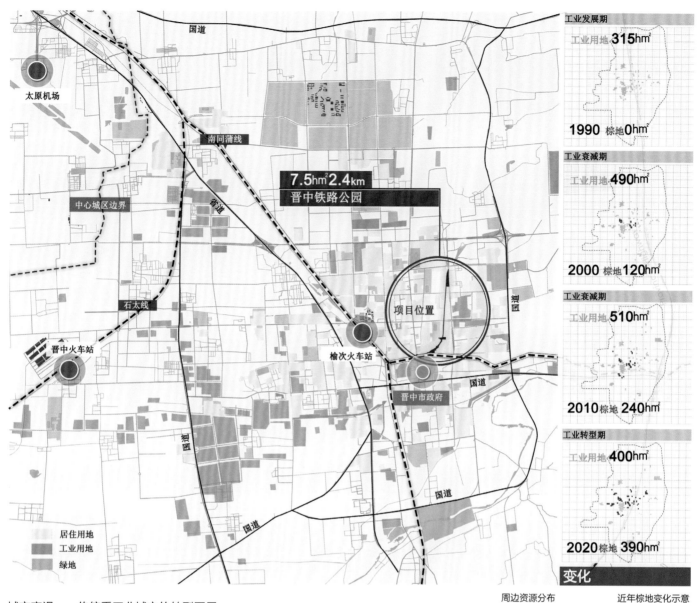

周边资源分布　　近年棕地变化示意

城市衰退——传统重工业城市的转型困局

晋中是中国传统工业城市，20 世纪 80 年代依托煤炭工业迅速崛起。城市的快速扩张和无序发展，带来了空气污染等生态问题。近年来，由于工厂的迁出，工业棕地的增多，晋中又面临着城市空间破碎化、社区环境老旧、铁路设施废弃等问题。在中国，与此类似的城市还有很多。衰退的工业城市正在努力争取转变，希望获得新生。

棕地——推动城市更新的核心潜力型困局

随着传统工业的萎缩和工厂外迁，煤运铁路被废弃。这条穿越城市的带状空间成为周边社区倾倒垃圾、排放废水的棕地。同时，失去了工厂的工业社区日益老化。低下的社区品质进一步加剧周边环境的恶化，形成恶性循环。据统计，该区域重度污染土壤达 5.8hm²，每年排入的雨水量近 4.5 万 m³，每年倒入的垃圾约 230t。规划初期，我们对晋中城区的工业棕地进行了详尽的调研。通过实地走访和多方交流，结合基于 ArcGIS 的空间大数据分析，评估了晋中工业棕地的空间潜力、生态潜力和文化潜力。准确地把控设计地块的综合功能和规划目标。

城市发展历史

场地现状分析

周边历史文化资源分析

- 城市热岛效应

139铁路位于城市的中心区域,周边热岛效应严重,急需缓解。

年平均热岛强度

0.4℃　　0.8℃　　1.1℃
1993—1996　1997—2006　2006—2020

自20世纪90年代以来,晋中市的城市热岛效应逐渐加重,30年间增长了180%。

:: Level 1:最弱热岛强度
:: Level 2:弱热岛强度
:: Level 3:强热岛强度
:: Level 4:最强热岛强度

- 空气质量

139铁路周边空气污染严重,有必要建设绿色通风走廊改善空气污染。

PM2.5	PM10	CO₂
51 μg/m³	126 μg/m³	2 mg/m³
NO₂	SO₂	O³
67 μg/m³	54 μg/m³	11 μg/m³

场址周围年平均空气质量指数约为110,比全市年均空气质量指数高28%。

:: Level 1:空气质量最高
:: Level 2:高空气质
:: Level 3:空气质量差
:: Level 4:空气质量最低

- 景观破碎度

139铁路周边景观破碎化程度较高,项目建设有利于提升城市整体生态连通性。

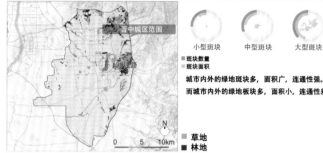

小型斑块　中型斑块　大型斑块
■斑块数量
■斑块面积

城市内外的绿地斑块多,面积广,连通性强。而城市内外的绿地板块多,面积小,连通性差。

■草地
■林地

- 生态适宜性

139铁路周边生态适宜性较低,伴有水、土壤和植被污染,需要进行生态修复。

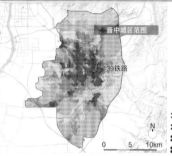

通过ArcGIS分析,确定多个评价指标,最终得出生态适宜性的区域分布。场地内生态适宜性高。

:: Level 1:最低生态适宜性
:: Level 2:生态适宜性低
:: Level 3:高生态适宜性
:: Level 4:最高生态适宜性

- 周边人口密度+年龄结构

139铁路周边人口密集,年龄分布广,对游憩的需求强烈,目前游憩空间不足。

0-14　　15-64　　65+

■2020 年龄结构
■2015 年龄结构

场地周围人口密度高。近年来,儿童和老人的比例逐渐增加。

:: Level 1:最低人口密度
:: Level 2:低人口密度
:: Level 3:高人口密度
:: Level 4:最高人口密度

- 社区的活力

该区域城市意向并不典型。大量的老旧设施和危房导致空间评价低。

环境活力
社会活力　→打分→ 社区活力价值
文化活力

建立社区活力评价体系,对环境活力、社会活力、文化活力进行综合评价。

● 高活力社区
● 一般活力社区
● 低活力社区

- 15分钟生活圈盲区

139铁路周围的居住区比较密集,但是绿地面积不足,所以需要增加大面积的绿地。

200m　500m　1000m
"5min" "15min" "30min"
步行距离

通过对15分钟生活圈覆盖范围的分析发现,由于周边居住社区密度较大,场地周边存在较强的游憩需求。

高游憩需求

低游憩需求　■绿色空间

- 可达性

139铁路周边交通便捷,但目前场地已经切断了周边交通,需要开放连接缝合。

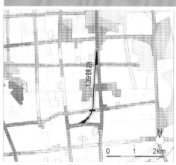

通过周边道路的可达性,分析区域开放空间的交通可达性。场地周边交通可达性高,有利于后期建设,方便游客。

:: Level 1:极低交通可达性
:: Level 2:低交通可达性
:: Level 3:高交通可达性
:: Level 4:极高交通可达性

场地资源评价

评估结果提示,139煤运铁路这条带状棕地的改建,对晋中市城市环境的整体更新有巨大的推动作用。其可能的价值包括:

1. 串联周边破碎的城市绿地和生活组团,构建城市绿色基础设施网络。
2. 改良土壤,构建水系统,营造生境,改善铁路及沿线生态环境,改善城市热岛效应,缓解空气污染。
3. 弥补周边社区缺失的游憩空间,改善社区环境,提升社区活力。

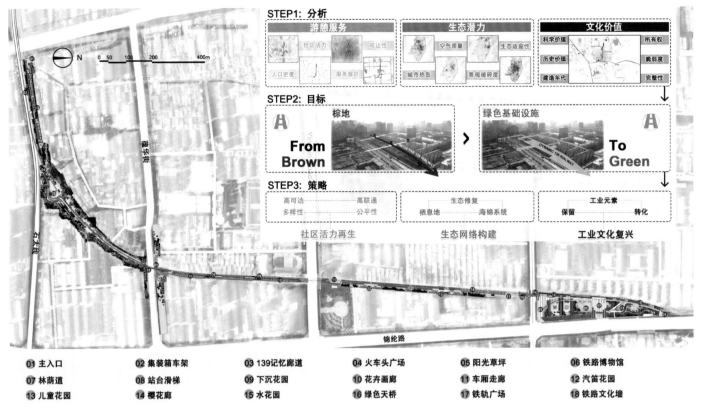

01 主入口	02 集装箱车架	03 139记忆廊道	04 火车头广场	05 阳光草坪	06 铁路博物馆
07 林荫道	08 站台滑梯	09 下沉花园	10 花卉画廊	11 车厢走廊	12 汽笛花园
13 儿童花园	14 樱花廊	15 水花园	16 绿色天桥	17 铁轨广场	18 铁路文化墙

从棕地到绿色基础设施——晋中铁路公园更新设计总平面

"更新"效果图

更新：基于社区居民需求，沿废弃铁路规划体育健身、儿童活动、生态科普、文化展览4大类型的16个社区绿色开放空间，构建连接周边社区新的活力中心。结合周边土地利用情况，合理布局绿色空间位置，形成联系7个社区的15分钟步行生活圈，确保绿色空间公平性。

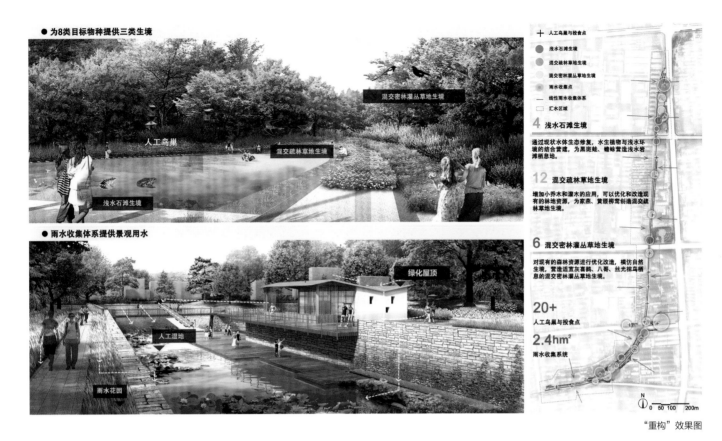

"重构"效果图

重构：规划通过一系列生态改造手段，将受到污染的棕地转变为绿色的生态廊道，包括垃圾清运填埋、土壤治理、植被修复和低影响开发。同时，通过生态系统修复和生境营造的方式，规划了22处生境组团、20处小动物喂食点，为晋中市内的8种本土动物提供栖息地和生存帮助。具体包括为鸟类迁徙和栖息提供帮助的混合林灌生境、林草生境，以及辅助两栖动物生存的湿塘石滩生境。

"再利用"效果图

再利用：设计对铁路遗存进行了景观再设计，139铁道公园将以全新的独特面貌展现该区域的历史文化和工业记忆。

6 城市更新与绿色开放空间　197

建成照片

6 城市更新与绿色开放空间 199

承载非物质文化遗产与生活的自然空间
——山西晋中社火公园规划设计

The Natural Space that Carries Intangible Cultural Heritage and Life
— Planning and Design of the Spring Festival Parade Culture Park, Jinzhong City, Shanxi Province

建成时间：2016年	Time of completion: 2016
建设地点：山西省晋中市榆次区	Construction site: Yuci District, Jinzhong City, Shanxi Province
项目面积：23.4 hm²	Project area: 23.4 hm²
建设单位：晋中市园林局	Construction unit: Jinzhong City Garden and Forestry Bureau

获 奖 信 息：2019年教育部优秀工程勘察设计园林景观设计一等奖
　　　　　　2019年中国勘察设计协会行业优秀勘察设计奖优秀园林景观设计二等奖
　　　　　　2019年国际风景园林师联合会亚太地区文化及城市景观类荣誉奖

Awards: First prize of Landscape Design for Excellent Engineering Survey and Design of Ministry of Education in 2019
Second prize in Excellent Landscape Design Award of Industry Excellent Survey and Design Award for China Survey and Design Association in 2019
Award International Federation of Landscape Architects Asia-Pacific Region Award (IFLA AAPME) Cultural and Urban Honorable Mention in 2019

主要设计人员：李　雄　戈晓宇　林辰松　李方正　刘利刚　等
Project team: LI Xiong, GE Xiaoyu, LIN Chensong, LI Fangzheng, LIU Ligang, et al

　　历史上千年的社火文化是中国丰富多彩的民间艺术的集合，2006年社火文化被国家列入第一批非物质文化遗产保护名录，体现了这一文化的重要性及保护这一文化的紧迫性。社火公园面积234000m²，以社火文化为主题，借助城市公园的力量为社火文化的保护与传承提供助力。

　　社火公园内部设置文化景观体系和雨洪管理体系，构建了自然与人工、现代与传统多元融合的高品质文化景观系统，为晋中多样的非物质文化提供了展示和活动的场所。同时它还是一个智慧弹性的基础设施，高效地实现了雨洪管理功能。通过将文化和多种功能与景观融合，社火公园成为晋中市一个兼顾社火文化传承保护、市民休闲活动承载、雨洪管理与调控的高品质城市文化公园。

With a history of thousands of years, the Spring Festival Parade culture is the collection of rich and colorful folk arts in China, in 2006, the Spring Festival Parade culture was listed in the first batch of intangible cultural heritage protection list by the state, which reflects the importance of this culture and the urgency of its protection. Covering an area of 234000 m², the Spring Festival Parade Culture Park takes the Spring Festival Parade culture as its theme and provides assistance for the protection and inheritance of the Spring Festival Parade culture with the help of the power of urban parks.

The Spring Festival Parade Culture Park set up the cultural landscape system with the stormwater management system, and constructs a high-quality cultural landscape system integrating nature and artificial, modern and traditional. It provides a place for the display and activities of various non-material culture of Jinzhong. It is also an intelligent and resilient infrastructure that efficiently implements stormwater management functions. By integrating culture and multiple functions with the landscape, the Spring Festival Parade Culture Park has become a high-quality urban cultural park in Jinzhong City that takes into account the inheritance and protection of Spring Festival Parade culture, the carrying capacity of citizens' leisure activities, and the management and regulation of rainwater.

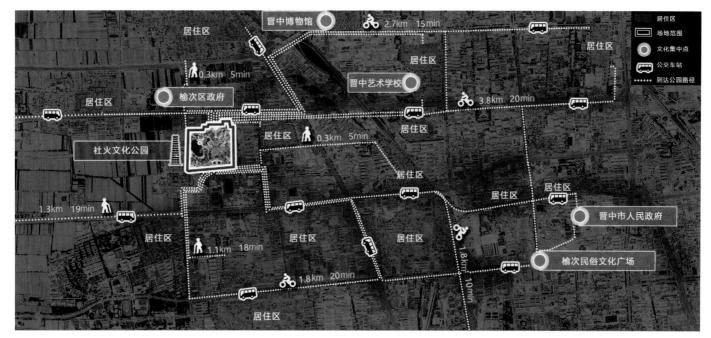

场地分析

公园建设时，晋中市公园绿地主要分布在城市中部及东部区域，场地周边稀缺大面积集中的绿地，以及供周边居民日常休闲游憩的公共空间。因此，社火公园的建设同时具有城市与乡村的双重民众基础，还是对晋中市绿地系统的有力补充。

设计团队一方面对晋中社火文化的艺术表现形式进行梳理及归纳，将其转译为生动活泼的雕塑布置到公园的各个活动场地，并对社火文化中最具特点的内容，如社日、架火塔等，通过形态的重塑、材质纹样的再现，凝练成景观构筑置于公园的重要节点中，展示文化的同时形成重要的景观控制节点；另一方面利用雨水花园及景观水体结合排水管道共同构成串联体系，雨水通过地形的塑造组织汇水，汇入雨水花园，当其达到最大蓄水量时，雨水将通过排水管道输送到下一个储蓄量更大的雨水花园，永久性景观水体作为雨洪管理体系的末端，将最大程度地消纳雨水，在形成优美水景观的同时，依靠径流收集形成季节性景观。

社火公园以景观作为核心驱动力，在满足绿地生态、游憩、安全等基本要求的情况下，形成了文化景观多元化、参与互动性强的绿色空间。

雨水花园建成实景

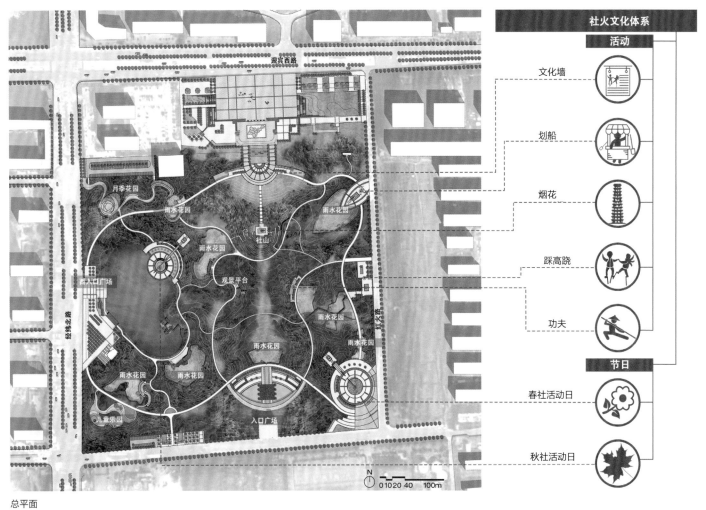

总平面

面对设计中应如何展现地域特征、展示社火文化、吸引游人融入体验，如何转变条件制约、消解自身内部径流、缓解城市径流威胁，如何巧妙利用现状，创造优美的景观、丰富市民的户外活动空间等多重挑战，设计团队提出应以景观作为核心驱动力，将文化景观体系和雨洪管理体系的构建串联起来，打造融合丰富文化内涵、优美地域景观、智慧雨洪设施于一体的城市公园。

202　理地营境　生态文明建设背景下风景园林实践

雨洪分析

　　场地位于晋中市榆次区，历年平均降雨量为490.6mm，夏季多雨且主要集中于7~8月，短时降雨量大。由于公园建设先于周边市政雨水管网建设，导致内源雨水无法及时外排。设计团队在园区内设置了8处雨水花园（13552m²）。通过雨水花园的分散布置，其汇水分区能基本覆盖整个园区。另外对场地内现存的鱼塘进行充分利用，将其改造为永久性景观水体（13276.9m²），雨水花园及景观水体结合排水管道共同构成串联体系。

建成实景

社火文化包括社日与社火活动。方案将社火文化广场与晋商文化广场相接，结合社火大事记墙记载晋中举办的各类社火文化活动，形成公园主轴线的北口开端；轴线制高点社山取祭祀土地神之意，山顶设有晋中传统社火塔，由钢结构打造，形成社火文化图腾与标志；轴线南端为社火博物馆，以覆土建筑形式与周边地形融为一体，为陈列社火文化历史等搭建了重要平台。社日层面，方案将春社区以节气为主题进行布局，设置春社文化景墙等载体，开展赏花祭社等春社活动；将秋社区分为秋社广场、光影长廊、秧歌广场城市亲水带、灯火主题广场，开展以赏月观灯、庆祝丰收为主题的秋社活动。社火活动层面，方案在系列文化主题节点以构筑、雕塑、技艺语言等方式表达了高跷、背棍、武技、秧歌等社火表演中最具乡土气息的形象，形成社火文化演绎的标志性景观，作为文化生活的聚集地。

社火文化的保护与传承

通过创造多级集雨汇水系统，场地已具备消纳不同暴雨等级的能力。根据 SWMM 分析结果，在降雨重现期为 2 年一遇（45.2mm）、3 年一遇（52.5mm）、5 年一遇（60.7mm）、10 年一遇（70.9mm）、20 年一遇（80.7mm）的条件下，社火公园对径流总量的削减率分别为 100%、65.9%、35.8%、24.3%、2.8%，对峰值流量的削减率分别为 100%、44.0%、31.2%、19.8%、8.7%，两年重现期峰值消失，与传统模式对比其峰值时间分别推迟 60min、32min、15min、5min。

雨洪管理体系构建的成功探索

社火公园的建设对晋中市绿地系统进行了有力补充，已经成为晋中市不可或缺的户外活动场所。公园内一系列文化主题场地为群众提供了丰富的活动场所，同时引入当地的晋剧票友协会、民间鼓乐队、秧歌社团、抖空竹健身协会等，满足市民的各类活动聚会，形成了鲜活的地域文化生活展示通道。

社火公园的建设突破了以往文化保护停留于单纯的展示，组织形式单一，无景观化的游览互动等的局限，为晋中市创造了一处传统与现代、城市与乡村交融的绿色空间，让文化遗产的传承与保护真正地融入市民的日常生活之中，同时通过雨洪管理体系的构建，场地已具备消纳不同暴雨等级的能力。社火公园的规划设计展示了如何以景观作为核心驱动力，在满足绿地生态、游憩、安全等基本要求的情况下，形成文化景观多元化、参与互动性强的绿色空间。

市民休闲游憩活动的焕活与承载

6 城市更新与绿色开放空间 205

活力的绿色开放空间
——青海西宁五一路口绿地广场设计

Vigorous Green Open Space
— Wuyi Road Intersection Green Square Design, Xining City, Qinghai Province

建成时间：2018 年	Time of completion: 2018
建设地点：青海省西宁市五一路	Construction site: Wuyi Road, Xining City, Qinghai Province
项目面积：4273 m²	Project area: 4273 m²
建设单位：西宁城辉建设投资有限公司	Construction unit: Xinning Chenghui Investment Co., Ltd

获奖信息：2018 年国际风景园林师联合会亚非中东地区经济价值类卓越奖
Awards: Award International Federation of Landscape Architects Asia-Africa Middle East Region Award (IFLA AAPME) Economic Value Excellence Award in 2018

主要设计人员：李雄 郑曦 戈晓宇 姚朋 等
Project team: LI Xiong, ZHENG Xi, GE Xiaoyu, YAO Peng, et al

　　五一路口空间的重新设计是西宁老城更新计划的成功项目，被认为是改善西宁城市中心交通状况、慢行环境和市民生活品质的战略举措。

　　在启动城市更新计划之前，西宁市旧城区现有的基础设施承载力趋于饱和，城市交通安全与拥堵问题随之萌生。与此同时，极高的住宅建筑密度与城市开放空间极度缺失形成突出矛盾，盲目的城市开发剥夺了城市居民安全出行与户外生活的权利，老城区街道失去秩序与活力，公共空间公平性低下的问题日益凸显。

　　项目处于西宁老城中心两条重要干道的十字路口，坐落在商业、教育、居住、政府办公等错综复杂建筑的聚集核心，是旧城片区开放空间失调矛盾的典型反映。风景园林师成为这一场地的协调者，综合考虑该地块承载的交通优化、城市功能疏解和街道品质提升的愿景之后，与开发部门等利益相关方达成多方共赢的共识，将原场地破败的店铺和国有企业建筑转移至更具发展潜力的城市新区，拓展车道缓解拥堵问题，对整体开放空间进行重新设计。由此，4000m² 的十字路口空间获得转型机遇。

The redesign of Wuyi Road entrance space is a successful project of Xining old city renewal plan, which is considered as a strategic measure to improve the traffic condition, slow traffic environment and the quality of life of citizens in the city center of Xining.

Before starting the urban renewal plan, the existing infrastructure carrying capacity of Xining old city tends to be saturated, resulting in urban traffic safety and congestion problems. At the same time, the extremely high residential building density and the extreme lack of open space in the city form a prominent contradiction, blind urban development deprived urban residents of the right to travel safely and outdoor life, the streets in the old city lose order and vitality, and the problem of low fairness of public space is increasingly prominent.

The project is located at the crossroads of two important roads in the center of Xining old city, and is located in the core of complex buildings such as commerce, education, residence and government offices. It is a typical reflection of the imbalance and contradiction of open space in the old city area. Landscape architect as a site coordinator, comprehensive consideration of the land carrying traffic optimization, the city function organization and improve quality of vision after the streets, and development departments and other stakeholders to reach a win-win consensus, the original sites and dilapidated shops and state-owned enterprises transferred to the new city district of greater development potential, expand the driveway to alleviate congestion, The whole open space was redesigned. Therefore, 4000m² intersection space has a transformation opportunity.

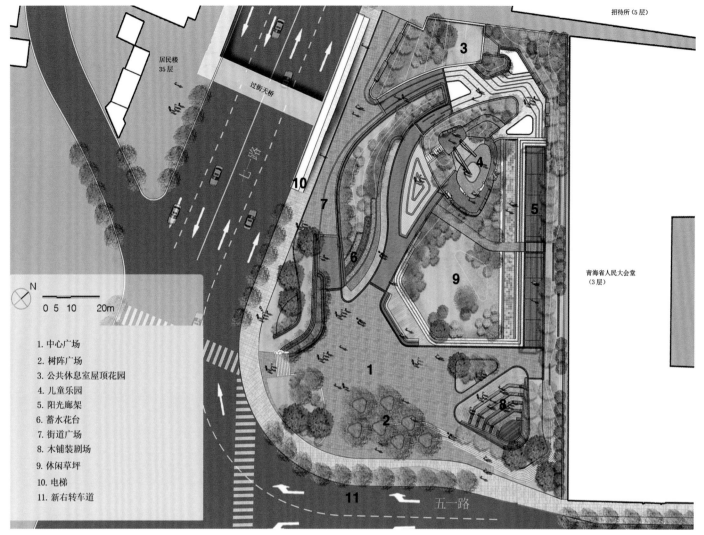

场地平面

五一路口开放空间在解决城市交通拥堵的基础问题的同时,将优质的公共空间回馈公共性、公平性低的街区,构建了一个包容的、复合功能的、可持续的城市枢纽。它系统地联系邻近的社区和主要街道慢行系统,为街区带来不可忽视的品质与价值提升。

高密度的开发挑战

6 城市更新与绿色开放空间

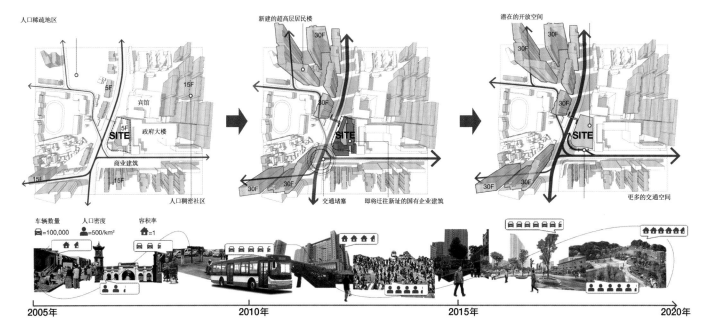

城市振兴的机遇

　　五一路十字路口空间在21世纪初曾经是西宁市区的繁华地区，被众多商业建筑和居民建筑环绕。2010年之前，西宁市还是西北地区的待开发区，从2010年超高层社区的建设启动开始，人口和机动车的数量经历了飞速增长，十字路口交通堵塞的问题日益恶化。到2015年，随着西宁城区更新计划启动，为了腾退出更多有价值的交通空间，十字路口东北部的原有建筑被迁往新址。

　　4273m²的绿色开放空间位于七一路和五一路交叉的十字路口。本项目基于交警部门的长期调研数据，提出扩宽右转车道的策略，优化了车行交通系统，绿色步行系统和人行道通过立交桥的连接与社区建立了更为密切的关系。

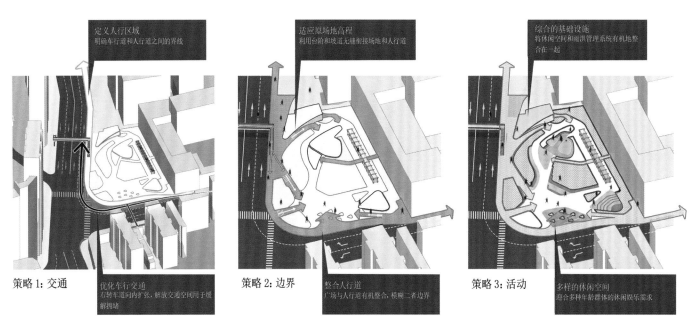

城市设计策略

　　设计消隐了场地与街道的边界，通过面向边界的街道广场使人行道设施与内部场地融为一体。无障碍坡道与台阶解决了场地与外部道路之间有不规则的3.5m高差，使广场内部与人行道竖向无缝衔接，并以种植池的耐候钢勾勒出坡道的边界合理回应人行天桥、斑马线等慢行系统要素，使场地内部路径融入城市慢行系统。乔木与低矮的挡墙形成通透的沿街立面，覆土建筑构成了广场与街道之间柔和的标志，广场为灰色且封闭的街道打开了绿色的窗口，欢迎所有市民进入、穿行与停留。

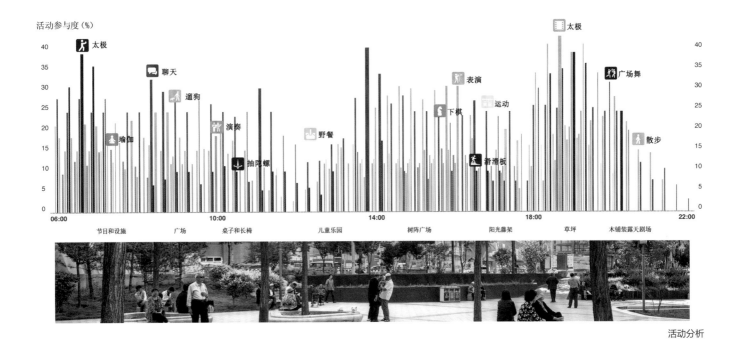

活动分析

该设计方案旨在创造新的城市公共空间——具有丰富的细节和视觉兴趣的、体现历史遗产价值的，与周围密集的城市肌理形成对比的，并使人心情舒畅的公共绿地。设计充分考虑了老城人群的年龄和族裔差异，布置应对多样性需求的公共空间，休闲草坪、小剧场、适合不同年龄段的儿童游戏场、阳光走廊和灵活使用的中心广场，满足市民预期的日常和季节性活动的需求。场地内的服务建筑被绿色屋顶覆盖，各种绿地空间都兼具雨洪调节设施的基本功能。

广场是通勤者的便捷通道，城市员工的避风港，以及由公共交通工具抵达的市民共同目的地；该处适合每一时段的日常使用，行人休息，家庭带孩子进行亲子活动，长者进行太极、抽陀螺、下象棋等活动；3类不同规格的街道家具提供了充足和舒适的座位，工人可以享用午餐，来自周边社区不同年龄和背景的城市居民可以分享和聚会。

丰富的社区设施

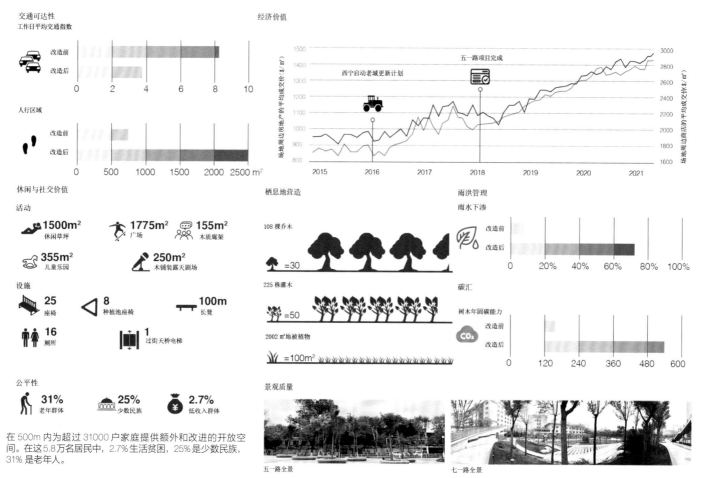

在 500m 内为超过 31000 户家庭提供额外和改进的开放空间。在这 5.8 万名居民中，2.7% 生活贫困，25% 是少数民族，31% 是老年人。

项目影响

无障碍坡道和台阶解决了场地和外部道路之间不规则的 2.5m 高差，改善了广场和周围城市肌理的所有游客的流通。

竖向连接

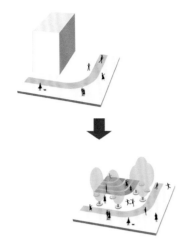

该项目提供各种包容性的公共空间,响应人口多样性,满足公众日常和季节性活动的需求。

包容的公共空间

本项目位于超高密度的旧城中心区,作为城市更新过程中蓬勃发展的绿色基础设施,创造了大量优质的公共空间,提高了绿化率,优化了街道可达性,促进了居民之间的交流。另外,本设计以海绵城市的概念为指导,作为弹性景观减弱了暴雨和干旱带来的负面影响。此外,乡土植物的魅力、乡土文化和地域景观的光芒也从这扇敞开的绿色窗户里透进来。

该项目为西宁的老城创造了一个具有卓越社会、生态和象征价值的目的地和聚集场所,一个当代设计的、可持续的、变革的城市景观,它为类似高密度开发环境下的项目提供了一种新的更新模式——一种谨慎平衡基础设施需求与公共空间权利的模式。

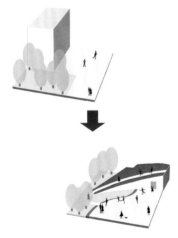

综合公用设施、历史展示墙和绿色屋顶建筑。娱乐场所和儿童游乐场与保留洼地紧密结合,形成复合的城市基础设施。

基础设施综合体

城市棕地中铁路遗迹的重生
——青海西宁通海桥公园

Rebirth of Railway Heritage in Urban Brownfield
— Tonghaiqiao Park, Xining Gity, Qinghai Province

建成/完成时间：2020 年	Time of completion: 2020
建设地点：青海省西宁市海湖新区	Construction site: Haihu New District, Xining City, Qinghai Province
项目面积：10.33 hm²	Project area: 10.33 hm²
建设单位：西宁城辉建设投资有限公司	Construction unit: Xining Chenghui Construction Investment Co., Ltd
获 奖 信 息：2021 年国际风景园林师联合会亚太地区建成项目类文化与城市景观类卓越奖 2021 年教育部优秀工程勘察设计园林景观设计二等奖 Award（IFLA AAPME）A 挖	Awards: Cultural and Urban Landscape Awards of Excellence in 2021 Second prize of Landscape Design for Excellent Engineering Survey and Design of Ministry of Education in 2021
主要设计人员：李 雄 郑 曦 戈晓宇 姚 朋 等	Project team: LI Xiong, ZHENG Xi, GE Xiaoyu, YAO Peng, et al

  青藏铁路是世界上最高的铁路，是连接中国西部青藏高原的天堂之路。作为青藏铁路的开端，兰青铁路在历史上发挥了重要作用。但随着城市的发展，这种独特的历史记忆正在渐渐消失。通海桥公园的场地作为兰青铁路遗址与城市新区矛盾最集中的地方，成为机遇与挑战并存的城市复兴目标。据此，我们提出了对于城市棕地的三种设计策略：①文化塑造；②空间重塑；③生态治理。场地内的铁路遗址在城市新区的发展中展现出新的活力，为了满足当代城市公共绿地的需求，设计利用地形对周边的交通噪声和雨水径流制定了有效的设计策略。该设计在铁路棕地的基础上有效地整合了铁路遗产记忆、周围居民的娱乐需求和区域生态系统的功能，承载了区域精神，并创造了一个具有吸引力的城市开放空间。

  基于文化塑造、空间重塑和生态治理这三种策略，兰青铁路已从交通设施棕地转变为具有文化特色，空间生命力和生态功能的新城市门户公共空间，从而唤起了人们对过去的记忆。

  The Qinghai-Tibet Railway, the highest railway in the world, is a paradise road connecting the Qinghai-Tibet Plateau in western China. As the beginning of qinghai-Tibet Railway, lan-Qing Railway played an important role in history. But with the development of the city, this unique historical memory is gradually disappearing. The site of Tonghaiqiao Park, as the place where the contradiction between the lanzhou-Qinghai Railway site and the new urban area is most concentrated, has become the target of urban rejuvenation with both opportunities and challenges. Accordingly, we put forward three design strategies for urban brownfield:①Cultural shaping;②Spatial remodeling;③Ecological governance. The railway site within the site has taken on a new vitality in the development of the new urban district. In order to meet the needs of contemporary urban public green Spaces, the design utilizes the topography to formulate effective design strategies for surrounding traffic noise and rainwater runoff. Based on the railway brownfield site, the design effectively integrates the memory of the railway heritage, the recreational needs of the surrounding residents and the functions of the regional ecosystem, carrying the regional spirit and creating an attractive urban open space.

  Based on the three strategies of cultural shaping, spatial remodeling and ecological governance, Lanzhou-Qinghai Railway has been transformed from a brownfield of transportation facilities into a new urban gateway public space with cultural characteristics, spatial vitality and ecological functions, thus evoking people's memories of the past.

1 入口广场
2 白桦林
3 观景台
4 紫丁香花园
5 草坪
6 火车模型
7 景天花园
8 集装箱画廊
9 平台广场
10 稀疏森林
11 娱乐广场
12 沿街的空地

场地平面

狭长的绿地为高密度市区提供了喘息的空间

6 城市更新与绿色开放空间 213

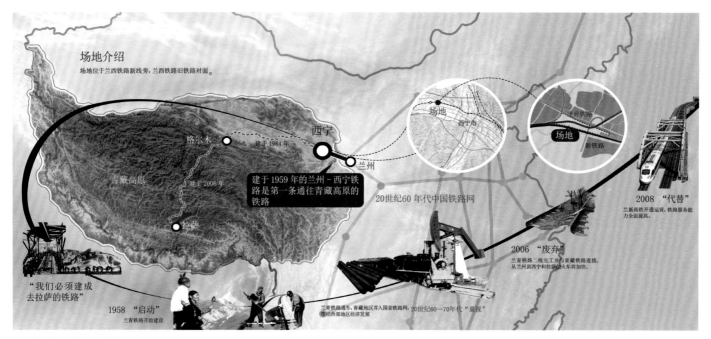

通往青藏高原的第一条铁路

1960年代，兰青铁路是青藏高原与外界连接的主要交通要道，具有很高的历史价值。兰青铁路的建设使13万人获得就业机会，改善了贫困地区的生产和生活条件，促进了青海的能源工业、金属加工业、农业和副食品加工业等的发展，是中国西部大开发的见证。

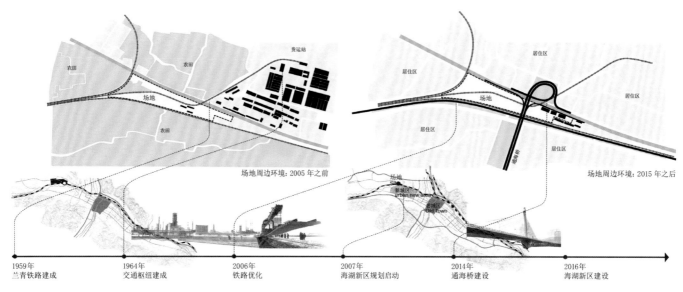

从交通枢纽到城市棕地

场地占地10.33hm²，靠近为应对城市扩张而建造的新城区。站点的北侧是主要道路，南侧是青藏铁路。是东西方向的一个长而狭窄的低洼地，被新建的居民区、道路和桥梁所环绕。穿过这片土地的兰西铁路通道已被废弃，中国西部的昔日辉煌在此已逐渐被人们遗忘，场地为了适应新市区的发展亟待更新。

现状条件与挑战

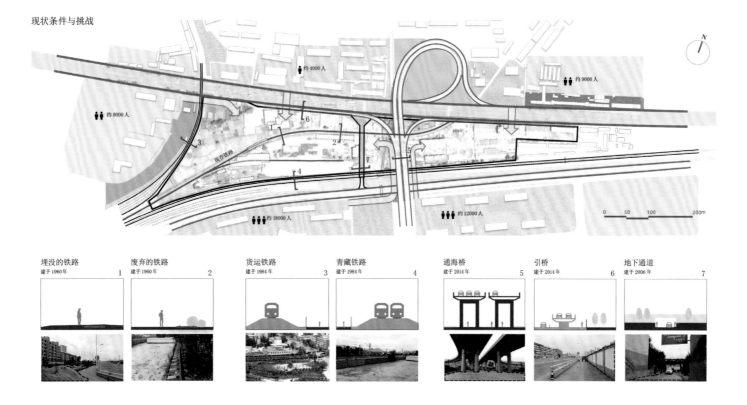

1. 现场废弃的铁路。作为重要的地方工业遗产，兰青铁路的废弃部分是延续城市记忆的重要载体。如何在公共场所保留和重新挖掘铁路的新含义是重新激发场地活力的首要任务。

2. 场地边界条件复杂。高度差和交通空间使站点与周围的人流分开。通海桥公园作为提高空间活力的公共空间，必须改善场地的通行性和步行便利性，并创建一系列令人兴奋的开口。

3. 毗邻公路、铁路和其他运输基础设施。噪声令人不快，地形上的洼地会从周围的道路和建筑物中积水。有效的降噪和雨水径流的整理是当代城市公共绿地的重要要求。

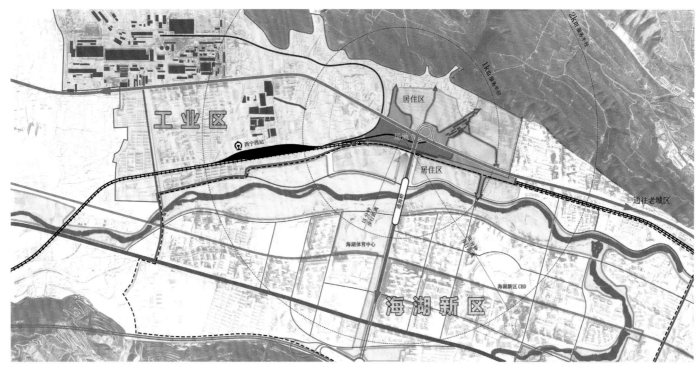

区域生态空间和开放空间的目标

设计方案应与规划的绿色空间相联系，将区域绿色系统的福祉渗透进周边居民和区域内市民的生活中。

策略一：铁路作为遗产与景观

现场的临时建筑和废弃工厂被拆除，将埋藏在地下的铁路遗址充分的展现。尽管这些设施失去了原有的功能，但它们为新设计提供了更多可能性。本设计结合了1960年代铁路设施的特点和场地现状，引入了功能性和趣味性设施。

从轨道布局的流线形式出发，在整个设计中提取平行、合并和分叉的语言作为园路的组织方法和形式基础。设计采用"平台"作为公园的特色景观。本设计在这些立柱的材料和基本布局的基础上进行提取和抽象，转译成为现代的钢框架。上述设计思路与公园的整体形象相结合，体现了火车站的景观特征。

场地与南侧的铁路之间设置了绿色的隔离带，防止人们在游览时误入铁轨而产生危险。

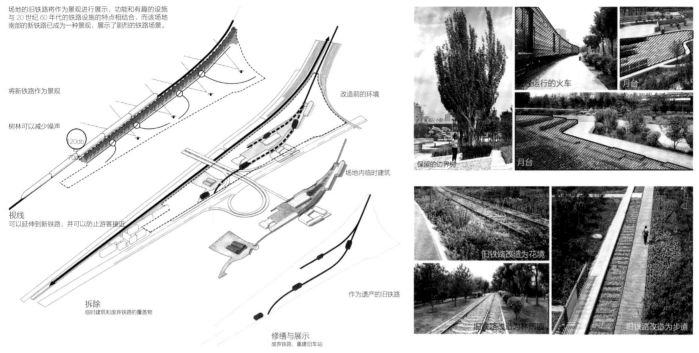

策略分析图一

策略二：铁路效益与功能规划

为了满足周边用户的无障碍需求和周边地区的使用需求，本设计将场地的边界转变为有趣的休闲和功能性设施，利用平台、轨道和其他的铁路设施的意向，场地边界成为具有铁路特色的休闲和功能区域。此外，高架桥引桥和工地内部新旧铁路高度差形成的空间也得到了积极的改造，成了一个积极的转折点，改善了空间利用率，增强了场地的亲人性。

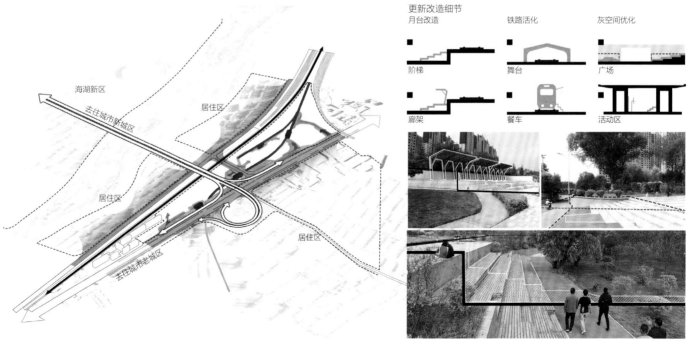

策略分析图二

策略三：雨水收集系统

该场地的中心部分受到市政道路和现有排洪沟的限制。该场地的内部 LID 系统设计如下：以泄洪沟和市政道路为界，在公园内处理场地东部和西部的雨水。当遇到极端天气现场的雨水无法及时处理时，多余的水将通过暗管排入市政雨水管网或当前的排洪沟。设计基于场地的当前地形，结合梯田和微观地形来组织排水系统，还使用了相关的 LID 措施，例如植草沟渠、生物滞留池、水净化植被台地、下凹绿地、可渗透铺装，以保留并初步净化和利用雨水。

当现场接收到的雨水达到阈值时，雨水将通过暗管排入市政道路的雨水管网。由于场地的原始地形是西部高而东部低，设计在现有的铁路和坝坡上种植草沟作为雨水传输设施，场地东部的雨水也通过微地形被排入植草沟，雨水从植草沟被引流至下凹绿地。

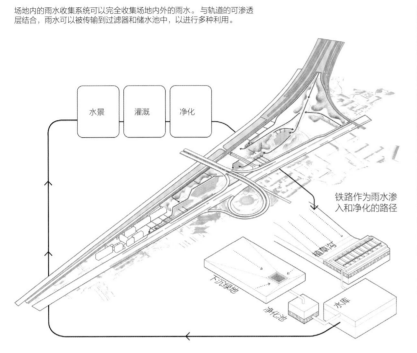

策略分析图三

项目效益

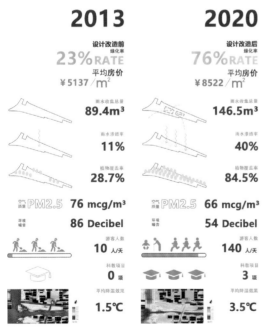

项目效益分析图

场地边界处理：不友好的边界高差被与功能形式相整合得到消解，被处理为具有停坐功能的场地

场地边界处理：场地内的旧火车成了多功能的室内空间

场地边界处理：场地内不友好的高差被处理为休闲娱乐空间

月台形制设计：保留了原场地内的白杨，铺装延续了铁路的特征

7

文化展园与艺术花园

Cultural Garden and
Art Garden

1987年第一届中国花卉博览会举办于北京全国农业展览馆，园林博览会正式进入国民公众的视野。1999年，昆明世界园艺博览会作为我国举办的第一个世界最高级别园林展，在国内外取得一致好评，也极大程度上促进了云南省乃至中西部地区旅游业和商贸业的发展。而后，随着我国经济的发展与生态文明建设理念的提出，我国园林博览会正式进入高速建设的发展时期，并迅速成为中国城市化进程中城市绿地系统完善、生态系统修复、经济外向发展、打造文化名片、建设基础设施的重要推手之一。

其中，城市展园类是目前中国园林博览会最为重要的组成部分之一，其核心目标为表述参展城市的历史文脉传承、自然风貌特征、人文精神典故等重要的城市地域文化符号。在相关研究统计中，城市展园数量占目前园林博览会中展园总数的60%以上。但高频次的园博园建设，导致部分城市展园的设计内容缺乏创新，简单地使用微缩景观或单调的文化复制展示，往往无法达到将参展城市地域文化名片进行科普展示的根本目的。因此，如何在城市展园有限的设计范围内形成基本空间骨架与叙事逻辑，如何合理选择表述城市地域特性的材料元素，如何将参展城市的文化符号进行展示与表达，成为城市展园设计的重要问题。

环境逻辑与空间逻辑

城市展园是园林博览会中的核心游赏展示区域，其空间逻辑既是设计师希望游客在城市展园内参与体验的基础叙事逻辑，也是决定整个城市展园概念陈述的基础逻辑。

线性连续的叙事空间逻辑是城市展园设计中常见的构成方式之一，通过对游客游园路线的引导与预估，在游线沿线的景观空间内将参展城市本土的人文历史、发展沿革、典籍传说等符号进行叙事化的展示与引导，使游客在完整的空间流线组织下以参与的方式进入城市展园的空间体验逻辑。线性的空间逻辑对递进逻辑的地域文化符号具有传递信息清晰、表述直观而明确的优势，对设计面积较大的城市展园容易形成空间环境起伏转合的变化关系，但也容易出现空间利用不充分、游赏路径不清晰可能带来体验并不完整的设计问题。

此外，以多个主题化空间对城市展园进行定义也是常见的空间构建逻辑，选择参展城市多个平行的地域文化符号与空间逻辑进行匹配，如精神文化、非物质遗产、特征地标等，以一个主空间加多个平行的子空间或单纯以多个平行子空间相互关联，从而形成组合式的城市展园空间构建逻辑。相较而言，组合式的空间逻辑适合同时对参展城市多项平行的地域文化符号进行传递和宣扬，对空间和游客游赏路线有着更强的自由度和更多样的体验感。

材料特征与氛围特征

在构建城市展园空间逻辑的基础上，需要对用来传递参展城市地域文化符号氛围特征的材料进行选择，这里的材料包括并不限于植物、山水、建构等元素。

植物材料作为城市展园景观设计中最重要的氛围营造材料之一，也是地域性文化特征表达的核心要素，通过对城市展园地域文化符号特征的提取，对应性地选择植物材料，采用片植或组合群落的不同种植形式，可以形成传达设计者不同思想主题的景观氛围。城市展园中的建构素材可以大致分为古今两类，以古典建筑或古建要素的设计为核心材料，选择参展城市明确的古迹名胜或某一特征类型的古建形式作为材料，明确直观地传递参展城市文化古迹、园林名胜等地域文化符号。以当今新材料、新技术为入手点，或师古创今，对传统文化进行转移与表述，或直言今意，对参展城市新时代发展做出阐述。此外，山水元素也是营造空间氛围的良好材料，提取参展城市自然环境特征，通过地形的开合收放、水体的亲疏远近，都可以营造不同的氛围特征。相对于植物和建构元素，山水元素对信息的传递更柔和隐晦。

文化载体与展陈载体

空间逻辑与氛围特征是城市展园地域性文化符号表述的基础，在此基础上需要对参展城市地域文化载体进一步阐述和说明，实现游客可观可赏的展陈载体，包括具备原型特征的具象化载体与取意境感受的抽象化载体。

将参展城市地域文化符号以实体形式进行展示与说明，包括直接缩移相应的建构载体、展示相应的物质符号、文字化或图示化相应的内容介绍等。对游客而言，这种具象化的表达方式使其对所传递的文化信息能够有着更清晰直观的认知与了解，对隐晦或复杂的文化内涵表述也更为简单明确。但部分文化载体更倾向于意识形态层面的感受体验，如人文精神、文化语境等，利用抽象化的符号载体，结合空间逻辑与氛围特征，游客能更身临其境地体会所传递的参展城市地域文化符号。

作为新时代城市生态文明建设的重要载体，园林博览会的举办，不仅是城市本身的文化展示与传播、新技术和新材料的应用实践，其中城市展园对地域性文化符号的传递更是参展城市宣传城市名片、推广城市文化、传承城市历史的重要途径之一。

本章选择了4个不同类型的城市展园，分别通过不同手法具体表现了其地方本土文化。

焦作七贤园探索与展示了博爱竹艺在现代风景园林中的应用方法。让人们从中了解和关注即将消失的手工艺文化遗产。

南阳五圣园以景观化的手法转译南阳乡土景观，由微缩景观的单纯堆砌转变为城市特征的艺术提取，打造城市的绿色缩影。

南阳邓州园从城市自身文脉特征出发，利用多维度感知的方法，从空间、视觉、嗅觉和触觉4个维度对应4种设计策略探索展园中展示城市文化的新路径。

保定清泽园是以新型低碳弹性材料"竹钢"组成结构单元，以水景空间为特色，以文化遗产展示为内涵的独特的城市展园景观，最终实现自然基底和人居理想空间的完美重合。

In 1987, the first China Flower Expo was held in Beijing National Agricultural Exhibition Center in 1987, and the garden expo entered the public vision for the first time. In 1999, Kunming International Horticultural Exposition, as the first top-level garden exhibition in the world held in China, won unanimous praise both at home and abroad, and greatly promoted the development of tourism, business and trade in Yunnan Province and even the central and western regions. Then, with the steady economic development and the proposal of ecological civilization, China's garden expo entered a period of rapid development and quickly became one of the most important driving forces for optimizing urban green space systems, restoring ecosystems, promoting export-oriented economy, building cultural landmarks, and developing infrastructures in the process of urbanization in China.

Among them, the urban exhibition garden is one of the most important components of China Garden Exposition at present. Its core goal is to express the important urban regional cultural symbols of the participating cities, such as the historical context inheritance, natural features, humanistic allusions and so on. According to research statistics, the number of urban exhibition gardens accounted for more than 60 percent of the total number of gardens in the current garden expositions. However, due to the high frequency of garden expo construction, the design content of some urban exhibition gardens is short of innovation, and the simple use of miniature landscape or monotonous cultural reproduction and display often fail to achieve the fundamental purpose of displaying and popularizing the regional cultural landmarks of the participating cities. Therefore, how to form the basic space framework and narrative logic within the limited design scope of urban exhibition gardens, how to reasonably choose the material elements to express the urban regional characteristics, and how to display and express the cultural symbols of the participating cities have become important issues in urban exhibition garden design.

Environmental logic and spatial logic

Urban exhibition garden is the core area for sightseeing and demonstration in garden expositions. Its spatial logic is not only the basic narrative logic that designers hope tourists to experience in urban exhibition gardens, but also the basic logic that determines the concept statement of the whole urban exhibition gardens.

The linear and continuous narrative space logic is one of the common forms in urban exhibition garden design. By guiding and predicting the tourist route, the local symbols of the participating cities, such as cultural heritage, development history, classics, and legends, are displayed and guided in a narrative way in the landscape space along the tour routes. It enables visitors to enter the spatial experience logic of urban exhibition garden in a way of participation under the complete spatial streamline organization. Linear space logic has the advantages of clear information transfer and accurate expression over regional cultural symbols of progressive logic. It is easier to realize environmental ups and downs for the urban exhibition gardens with large area but may also lead to incomplete experience due to inadequate space utilization and unclear tourist routes.

In addition, it is a common spatial construction logic to define urban exhibition gardens through multiple themed space, by matching multiple parallel regional cultural symbols and spatial logic, such as spiritual culture, non-material heritage, feature landmarks etc., with a main space with multiple parallel subspaces or purely through the correlation of multiple parallel subspaces to form a combined urban exhibition garden space construction logic. In contrast, the combined spatial logic is suitable for the simultaneous transmission and promotion of multiple parallel regional cultural symbols of the participating cities, with more freedom and more diverse experience for the space and tourist routes.

Material characteristics and atmosphere characteristics

Based on constructing the spatial logic of the urban exhibition garden, it is necessary to choose the materials used to convey the characteristics of the regional cultural symbols of the participating cities. The materials here include but not limited to plants, landscapes, and structures.

As one of the most important atmosphere construction materials in urban exhibition garden landscape design, plant materials are also the core elements of regional cultural features expression. By extracting the regional and cultural symbols of the urban exhibition gardens, the designers can convey different themes and landscape atmospheres through plate plants or combined community of different plant species. The construction materials in urban exhibition gardens can be roughly divided into two categories: ancient and modern. For the ancient style, the design of classical architecture or ancient architecture elements is taken as the core material, and the specific historical sites or ancient architecture forms of a certain characteristic type in the participating cities are selected as the materials to convey regional cultural symbols such as cultural relics, gardens and scenic spots of the participating cities clearly and intuitively. The modern type, however, either conveys the traditional culture through new materials and technologies or directly expounds the latest development of the participating cities in the new era. In addition, the landscape element is also a good material to create a space atmosphere. By extracting the characteristics of the natural environment of the participating cities, different atmosphere characteristics can be created through the opening and closing of the terrain and the proximity and distance of the water. Compared with plants and construction elements, landscape elements convey information in a softer and subtler way.

Cultural medium and exhibition medium

Spatial logic and atmosphere characteristics are the basis for the expression of regional cultural symbols in urban exhibition gardens. On this basis, further elaboration, and explanation of regional cultural carriers of the participating cities are needed to realize the exhibition carriers that tourists can feel and appreciate, including the concrete carriers with prototype characteristics and the abstract carriers with artistic conception feelings.

The regional cultural symbols of the participating cities will be displayed and explained in the form of entity, including direct contraction of the corresponding construction carrier, display of the corresponding material symbols, textual or graphic introduction of the corresponding content and so on. For tourists, such a concrete way of expression enables them to have a clearer and more intuitive cognition and understanding of the cultural information conveyed, and to express obscure or complex cultural connotations more simply and clearly. However, some cultural carriers are more dedicated to the experience at the ideological level, such as humanistic spirit and cultural context. Through abstract symbol carriers combined with spatial logic and atmosphere characteristics, tourists may have more immersive experience of the regional cultural symbols of the participating cities.

As an important carrier of urban ecological civilization construction in the new era, garden expos are not only the demonstration of the city culture and application of new technology and materials, but also an important approach to convey the regional cultural symbols, publicize the urban landmarks, promote urban culture, and inherit the urban history.

In this chapter, four urban exhibition gardens in different types are selected to express their local culture in different ways.

Jiaozuo Seven Sages Garden explores and demonstrates the application of bamboo art in modern landscape architecture, so that people may have a better understanding of and pay attention to the disappearing handicraft heritage.

The Five Sacred Garden in Nanyang transforms the local landscape of Nanyang into the artistic extraction of urban characteristics from the simple piling up of miniature landscape, and thus creates a green miniature of the city.

Starting from the characteristics of the city's cultural context, the Dengzhou Garden uses the method of multi-dimensional perception to explore a new way to display urban culture in the exhibition garden from four dimensions of space, vision, smell and touch in response to four design strategies.

The Baoding Qingze Garden is composed of a new type of low-carbon elastic material "bamboo steel" structural unit, featuring waterscape space and cultural heritage display as the connotation of the unique urban garden landscape, and finally realizes the perfect combination of natural base and ideal human settlement space.

基于文化遗产复兴的设计实践
——河南焦作七贤园

The Renaissance of Handicraft's Cultural Heritage in Landscape Architecture
— Seven Sages Garden, Jiaozuo City, Henan Province

建成时间：	2017 年	Time of completion:	2017
建设地点：	河南省郑州市航空港经济综合实验区	Construction site:	Airport Economy Zone, Zhengzhou City, Henan Province
项目面积：	1453 m²	Project area:	1453 m²
建设单位：	焦作市园林绿化管理局	Construction unit:	Jiaozuo Landscaping Administration

获奖信息：2018 年国际风景园林师联合会亚非中东地区文化与传统类杰出奖
2019 年行业优秀勘察设计奖优秀园林景观设计二等奖
2019 年教育部优秀工程勘察设计园林景观设计一等奖
Awards: Award International Federation of Landscape Architects Asia-Africa Middle East Region
Award (IFLA AAPME) Culture and Traditions Outstanding Award in 2018
Second prize of Excellent Landscape Design for Industry Excellent Survey and Design Award in 2019
First prize of Landscape Design for Excellent Engineering Survey and Design of Ministry of Education in 2019

主要设计人员：李　雄　肖　遥　张云路　戈晓宇　林辰松　等
Project team: LI Xiong, XIAO Yao, ZHANG Yunlu, GE Xiaoyu, LIN Chensong, et al

　　焦作七贤园是第十一届中国（郑州）园林博览会焦作市艺术展园，面积 1453m²，位于园博会西翼。该项目探索与展示了博爱竹艺在现代风景园林中的应用方法，让人们从中了解和关注即将消失的手工艺文化遗产。

　　有别于传统设计—施工的运行模式，该项目将参数化非线性景观设计与手工竹编相互配合，利用现代手段与传统技艺共同指导设计。同时，在施工过程中融合了焦作博爱竹编的传授、展示和推广应用，有效降低了工程成本，并起到了保护传承竹编工艺文化遗产的作用。

　　项目借助中国博览会的平台，将风景园林带动遗产保护设计模式进行了广泛推广，带来了直接的社会效益，对未来竹艺发展产生了深刻、长远的影响，为社会各界提供了手工艺遗产保护与文化创新的新思路。

Seven Sages Garden of the 11th China International Garden Expo covers an area of 1453 m². This project explores and demonstrates the application of Boai Bamboo Arts in the modern landscape so that people may understand and care more about this disappearing handicrafts and cultural heritage.

Different from the traditional design – construct operation model, the project will make sure parametric non-linear landscape design and hand-weaving skills cooperate with each other, using modern methods and traditional techniques together to guide the image design. In the process of construction, the project integrated Jiaozuo Boai Bamboo's teaching, display and promotion, which effectively reduces the project cost and plays a big role in protecting and inheriting the bamboo cultural heritage.

With the help of the platform of China Expo, the project promoted the design mode of heritage conservation driven by landscape architectures, and produced direct social benefits. It has a profound and long-term influence on the future development of bamboo arts and provides new ideas of handicraft protection and cultural innovation for all sectors of the society.

竹编现状

具有综合社会效益的园林工程——示范、培训、传承

设计使用了村民参与的施工方式。该过程分为五个阶段：第一阶段，设计师依托博爱县政府，对掌握竹编技艺并决定从事该竹艺项目的竹匠进行普查，并成立竹艺协会；第二阶段，协会向村民提供免费教学；第三阶段是组织学员按照设计需求进行编制，选拔优秀作品并回购；第四阶段，施工方聘请有经验的竹匠建造竹框架，并安装由学员制作的竹编表皮；第五阶段，组织学员到工地参观施工过程并学习竹编在园林中的应用方式。

新的项目模式实现了项目团队和村民的双赢。对项目本身来说，节省了成本，提高了竹编的准确性。项目取得了良好的景观效果，展示了具有地域特色的传统文化。对于村民来说，该项目为当地创收提供了有效的途径。更重要的是，整个过程让村民看到了竹艺在一个新领域的应用，可以激励他们主动探索新的工艺和产品。

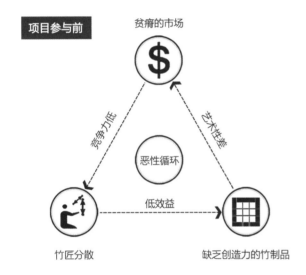
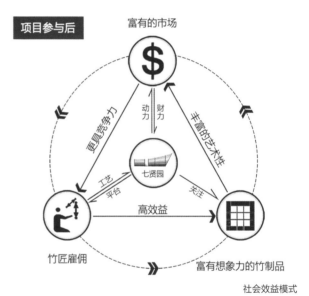

社会效益模式

焦作七贤园建成实景

现代手段和传统技艺的结合——参数化非线性景观设计与手工竹编

竹编结构复杂，形式多样，本项目设计使用了参数化软件进行建模。在数字工具的支持下，设计师可以考虑非线性景观的空间形态，模拟各种孔隙结构，推演光影变化，为手工竹编的表皮效果选择最佳的解决方案。该项目将使参数化非线性景观设计与手工竹编相互配合，运用现代手段与传统技艺共同指导设计。

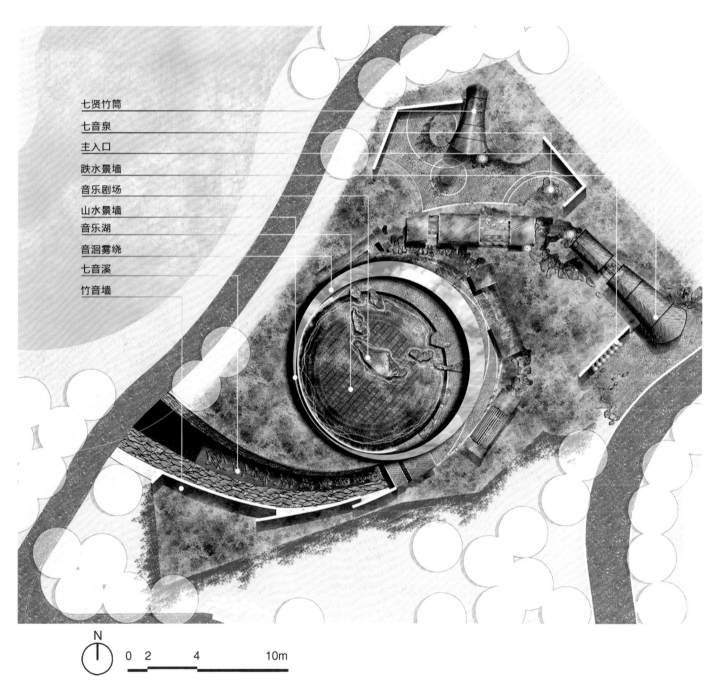

七贤竹筒
七音泉
主入口
跌水景墙
音乐剧场
山水景墙
音乐湖
音涧雾绕
七音溪
竹音墙

总平面

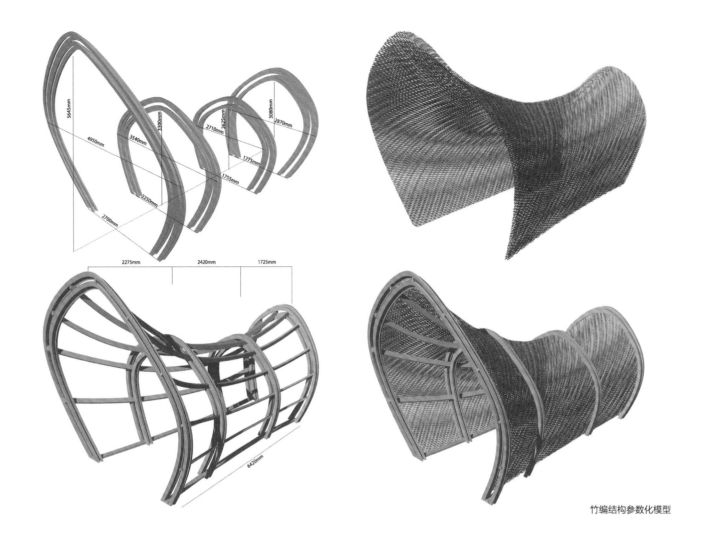

竹编结构参数化模型

以园博会为平台——风景园林带动遗产保护设计模式的推广

第十一届中国国际园林博览会吸引了近 500 万的游客。焦作七贤园的设计内容直观地展示了当地的传统文化底蕴，对城市形象的提升起着重要的作用。同时，借助园博会的平台，项目将风景园林带动遗产保护设计的模式进行了广泛推广，带来了直接的社会效益。博爱县目前已成立竹艺协会，会员 70 人，他们将继续从事竹编事业。项目设计团队还在博爱县周边 3 个乡镇的 20~60 岁人群中开展了竹文化的推广和普及活动。通过乡镇政府的反馈，广大村民对竹文化有了全新的认识，表达了学习竹编技艺的意愿，这对未来竹艺发展会产生深刻、长远的影响。

| 编织竹筒样品 | 切割竹竿 | 编织竹编 |
| 现场组装 | 维护管理 | 交流学习 |

构筑木骨架是将慈竹与树脂复合，经热压胶合制成的竹板，并加工成各种形式。这是一次将快速生长的竹子进行产业化加工的尝试。竹筒的表皮设计以焦作博爱传统竹编图案为基础。这些表皮是由当地的竹编工匠手工制作的，呈现出不同的光影变化。

竹编结构建成实景

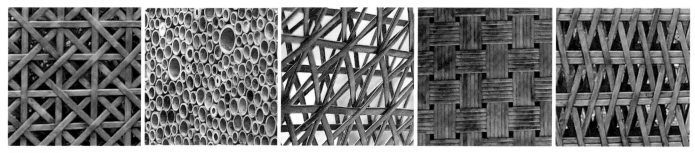

传统竹编图案

竹音墙：所有游客在入园时都可以得到一把小竹锤，敲击竹音墙，体验古时的雅致乐趣

休息空间：充满着东方禅意，这个竹林中的空间，无论是落水，还是竹声的回响，都让人心旷神怡

7 文化展园与艺术花园 231

竹制构筑物：这几处竹制构筑物是艺术展园的核心，它继承了中国传统风景园林"步移景异"的特点

竹音墙：池底刻有魏晋"七音"，沿水系挂在墙上的竹筒可以进行敲击，发出乐声

城市地域文化的多样性表达
——河南南阳五圣园
Diversified Expression of Urban Regional Culture
— Five Saints Garden, Nanyang City, Henan Province

建成时间：2017 年	Time of completion: 2017
建设地点：河南省郑州市航空港经济综合实验区	Construction site: Airport Economy Zone, Zhengzhou City, Henan Province
项目面积：1514 m²	Project area: 1514 m²
建设单位：南阳市园林绿化管理局	Construction unit: Nanyang Landscaping Administration
获 奖 信 息：2019 年行业优秀勘察设计奖优秀园林景观设计二等奖 2019 年教育部优秀工程勘察设计园林景观设计二等奖	Awards: Second prize of Excellent Landscape Design for Industry Excellent Survey and Design Award in 2019 Second prize of Landscape Design for Excellent Engineering Survey and Design of Ministry of Education in 2019
主要设计人员：李 雄 肖 遥 张云路 戈晓宇 林辰松 等	Project team: LI Xiong, XIAO Yao, ZHANG Yunlu, GE Xiaoyu, LIN Chensong, et al

外国园林展中，艺术展园主要以家庭园艺花园、主题花园、公共艺术展览、园林材料和设施展览为主。在中国，自昆明世界园艺博览会开始城市艺术展园之先河，城市艺术展园逐渐发展成为中国园林展的重要展示部分。历经多年多次的展会，城市艺术展园的设计手法和景观面貌开始趋同。如何立足参展城市自身特色，实现艺术展园设计的创新，是城市艺术展园设计的新挑战。设计团队希望通过本次实践，探讨城市艺术展园地域性文化的多样化表达手法。

中国国际园林博览会是国内园林花卉界最高层次的盛会，第十一届中国国际园林博览会的办会地点位于郑州航空港经济综合实验区，打造了园博会主园区和苑陵故城遗址公园两大园区，形成古今对话，象征历史文化的传承与发展。南阳艺术展园位于华中地区城市艺术展园区，占地 1514m²。设计以景观化的手法转译南阳乡土景观，由微缩景观的单纯堆砌转变为城市特征的艺术提取，打造城市的绿色缩影。

In international garden Expos, art exhibition gardens mainly include family garden, theme garden, public art exhibition, and exhibition of garden materials and facilities. In China, since urban art exhibition garden appeared in Expo of Kunming, it has gradually developed into an important part of Chinese garden exhibition. After many years, the design techniques and landscape features of urban art exhibition garden began to converge. The design of urban art exhibition garden is faced with a new challenge, that is, how to base on the characteristics of participating cities and realize the innovation. Through this practice, the design team hopes to explore the diversified expression of regional culture in urban art exhibition garden.

China International Garden Expo is the highest level event in domestic garden and flower industry. The 11th China International Garden Expo is located in Zhengzhou Airport Economy Zone. It has created two parks, the main park of Expo and the Yuanling Ruins Park. Forming a dialogue between the ancient and the modern, it symbolizes the inheritance and development of history and culture. With an area of 1514 m², the Garden of Nanyang is located in the Urban Art Exhibition Park of Central China. Through artistic extraction of urban characteristics, rather than simple stacking of miniature landscape, this design translates local landscape and creates a green miniature of Nanyang.

设计主题——"忆城·惜花·传文"

项目基于乡土地域性特征提取的主题探究，区别于使用微缩城市代表性景观的常规设计手法，我们希望着重挖掘参展地的地域特征及文化内涵，以统一、直观的景观转译手法，将其"重组再生"。南阳艺术展园以"忆城·惜花·传文"为题，将南阳的悠久历史与风貌之美缩于方寸之间。

利用空间形态表现地域性特征

夯土墙构架了南阳艺术展园主要的空间骨架。将汉代宛城及其建筑空间等的转化，利用古朴的夯土墙体，结合地形、种植等景观元素，形成汉宫门主出入口、月季历史甬道、下沉剧场以及汉宫墙苑等，模拟南阳都城的空间特征。

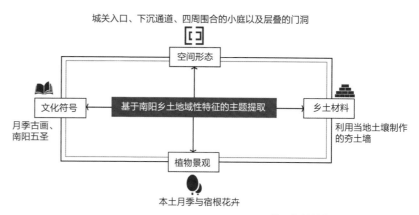

基于乡土地域性特征的景观转译模式

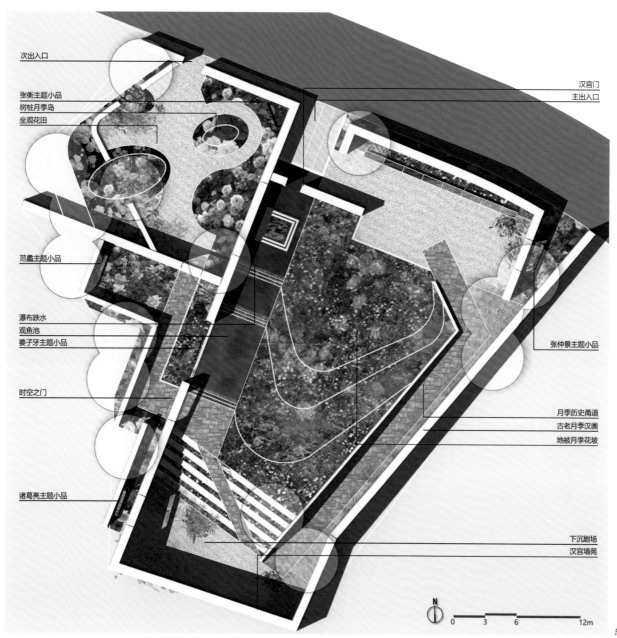

总平面

模板支设　　　　　　捆扎模板　　　　　　土墙夯筑　　　　　　模板拆卸　　　　　　土料抹平

利用材料表现地域性特征

艺术展园以夯土墙为骨架。现代夯土墙建造技艺源于古法，所用材料为会展当地土壤，依土样特性，辅以人为调色加工，形成自然而独特的夯土色彩和墙面肌理，且不产生任何建筑垃圾。运用会展地乡土材料构筑艺术展园风物，是一种乡土景观的再生。墙色源于大地，肌理独一无二，经时间洗礼之后的夯土墙表面会产生轻微风蚀，更添古朴韵味。

夯土墙

南阳五圣园建成实景

利用植物景观表现地域性特征

　　为呈现最本真的乡土景观,南阳艺术展园以本土月季为主要景观植物进行设计。地被月季和宿根花卉混植而成的花坡、点缀园中各处的各色月季组团、次入口的树状月季区域和水中的古桩月季,无一不是南阳月季丰富种类最直观的展示。既为达到艺术展园在会展期间的最优景观效果,又要使全年周期内有景可观,选择花期跨度较大的月季品种交替种植,延长主景月季的观赏时长。还运用一些创新的植物搭配方法,可作为新的月季植物景观搭配模式的探索。

月季和宿根花卉混植

利用文化符号表现地域性特征

月季和南阳五圣是本次南阳艺术展园最主要希望表达的文化内容。将古老月季品种作为墙面装饰灯带，在下沉庭院绘制观花古画，追溯汉代赏花之情境。五圣也以或雕或镂的形式隐藏在园中。

文化符号

南阳五圣园建成实景

唤醒城市记忆
——2019 世界月季洲际大会河南邓州园
Bring Back the City's Historical Memory
— Dengzhou Garden of 2019 WFRS Rose Regional Convention, Henan Province

建成时间：2019 年	Time of completion: 2019
建设地点：河南省南阳市卧龙城区	Construction site: Wancheng District, Nanyang City, Henan Province
项目面积：1377.6 m²	Project area: 1377.6 m²
建设单位：邓州市自然资源和规划局	Construction unit: Dengzhou Natural Resources and Planning Bureau

获奖信息：2019 年世界月季洲际大会城市展园设计金奖
2021 年国际风景园林师联合会亚太地区文化及城市景观类荣誉奖
Awards: 2019 World Rose Intercontinental Conference City Exhibition Park Design Gold Award
Award International Federation of Landscape Architects Asia-Pacific Region
Award (IFLA AAPME) Cultural and Urban Honorable Mention in 2021

主要设计人员：李　雄　林辰松　张云路　肖　遥　等
Project team: LI Xiong, LIN Chensong, ZHANG Yunlu, XIAO Yao, et al

　　作为河南重要的历史文化名城，邓州正面临着传统城市文化逐渐消逝的威胁。由于 2019 年南阳世界月季洲际大会的举办，其中设置多个城市展园，让邓州拥有了对外展示城市文化的绝佳机遇。本项目从城市自身文脉特征出发，利用多维度感知的方法，从空间、视觉、嗅觉和触觉 4 个维度对应 4 种设计策略探索展园中展示城市文化的新路径。设计利用城市独有的结构特征塑造展园的空间基底，利用地域代表性景观作为展园的重要展示节点，利用城市乡土植物营造展园的植物景观，利用城市传统的建造材料作为展园内各类构筑和铺装的主要材质。通过空间形态的转译、地域景观的提取、植物景观的营造和乡土材料的应用四大策略，唤醒邓州的历史记忆，延续城市传统文脉，展现文化底蕴。

　　该项目将创新的设计策略进行实地探索，与业主、施工方和主办方紧密协作，在短时间内以预算的造价完成了设计和施工，实现了良好的建成效果。获得了中国花卉协会和主办方一同颁发的设计金奖，在大会展园中具有极佳的示范性，为未来城市展园的设计提供了一种可推广的新方法。

As an important historical and cultural city in Henan Province, Dengzhou is facing the threat of the gradual disappearance of its traditional culture. Thanks to 2019 WFRS Rose Regional Convention, which sets up a number of city expo gardens, Dengzhou has seized an excellent opportunity to display its urban culture to the whole world. Based on Dengzhou's unique characteristics, this project uses multixdimensional perception methods to explore new paths for displaying urban culture in the expo garden from the four dimensions of space, vision, smell and touch. The four dimensions corresponding to four design strategies: 1) using unique structural characteristic of the city to shape the space of Dengzhou Garden; 2) using regional landscape as important display node; 3) using native plant to create flora landscape; 4) using traditional construction materials as the main materials for various structures and paving in the garden. Namely, through translation of space, extraction of regional landscape, selection of plants and application of local materials, design can successfully arouse people's memory of Dengzhou, and achieve the desired effect of continuing its traditional context and displaying cultural heritage.

This project applied these innovative design strategies to practice. The design team worked closely with the owner, constructor and organizer to complete the task from design to construction within a short period of time at the budgeted cost. Finally, Dengzhou Garden has achieved a good completion effect. It has also won the gold award of design issued by the China Flower Association and the organizer. Dengzhou Garden has been an excellent demonstration among city exhibition gardens and provided a new method which can be promoted for the design of city exhibition garden in the future.

场地平面

设计方案

项目地点位于中国河南省南阳市，2019年南阳世界月季洲际大会中的城市展园区，总占地面积1377.6m²。设计从多个维度对邓州进行展示，共设计了10个景观节点，实现了在有限的用地内最大程度展现城市历史文脉，游客可以通过空间感、视觉、嗅觉和触觉多个维度进行体验。在游览过程中强化游客对于城市特殊结构的空间感知力度，拓宽城市地域景观的表达方法，触摸邓州古老的城市记忆，延续城市传统文脉。

设计目标

1. 变微缩景观为特色凝练，展现城市深厚历史；
2. 变走马观花为深度体验，多维度感知邓州文化；
3. 变普世通用为乡土情怀，传递城市特质与内涵。

设计理念

通常，人们借助多种感官而不是纯粹的视觉来体验城市，包括听觉、触觉、嗅觉乃至味觉。而如今的展园设计却缺少对人的多维度感知的关注设计。从邓州自身的城市文脉特征出发，我们选择从空间、视觉、嗅觉和触觉这4个维度进行本次展园的设计，巧妙地利用了多维度感知的优势展示城市文化。

空间感：利用城市独有的结构特征塑造展园的空间基底。

视觉：利用地域代表性景观作为展园的重要展示节点。

嗅觉：利用城市乡土植物营造展园的植物景观。

触觉：利用城市传统的建造材料作为展园内各类构筑和铺装的主要材质。

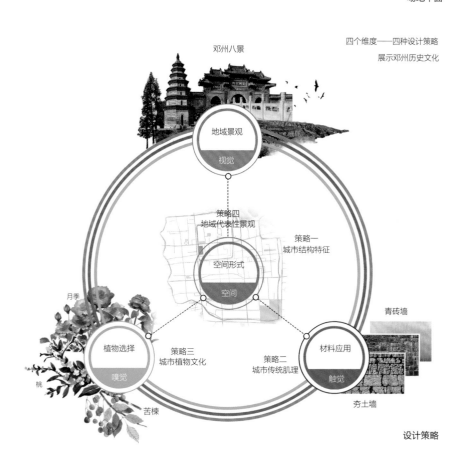

设计策略

基于四维感知的方案解读

1. 空间感：空间形态的转译

基于邓州城市发展过程中最具代表性的城市空间特征——双重城墙，以及自然空间特征——花洲，将展园的空间基底划分为两部分。

双重城墙空间：将古代邓州内城墙、外城墙和护城河的城市结构进行了现代转译，通过错落的景墙以不同的围合方式，塑造出交通空间、停留空间和展示空间。

花洲空间：将邓州传统花洲形态转译为展园内主要的植物展示空间，分别利用灰色和白色砾石代表水域和浅滩，通过流畅的铺装曲线划分出多个种植岛。

2. 触觉：乡土材料的应用

邓州传统双重城墙中，外墙为夯土结构，内墙为砖结构，传统夯土墙利用乡土材料建造，墙面质地细腻，但建造过程复杂。设计时对传统夯土墙进行了改进，保留了夯土和青砖两种材质，选择了特制的仿夯土混凝土板，利用现代工艺还原古城质感。建造完成后，游客触摸着象征着内外城墙的夯土与青砖肌理，逐渐走入城墙空间中，感受着古城最初的历史文脉。

3. 嗅觉：植物景观的营造

植物景观分为城市主题植物、花洲主题植物以及大会主题植物三部分，利用不同植物的气味还原城市记忆。城市主题植物选择苦楝、白榆等乔木，呼应邓州古城墙的植被现状；花洲主题植物选择圆柏、梧桐、桃树和李树，还原花洲色彩明快、活泼的植物景观氛围；大会主题植物选择多个品种的月季，再现百花洲景观风貌，烘托出浓烈的盛会氛围。

4. 视觉：地域景观的提取

邓州古八景历史悠久，后随着历史的变迁与环境的改变，直至清乾隆年间形成稳定的八景体系。展园以《邓州八景诗》为蓝本，从画、名、诗三方面全方位展现邓州古八景。八景画雕刻在平整的石面上，布置于双城空间的护城河水景内，池壁雕刻有八景名，八景名背后的景墙刻有八景诗。最终形成八景画、名、诗的景观序列，景画藏于水中，沿池壁而上即见景名，景名之后是因高差形成的流水诗卷，共同组成完整的邓州八景地域景观视觉展示体系。

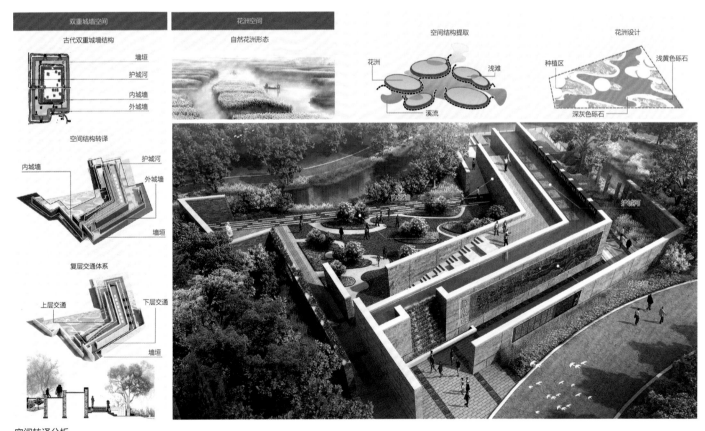

空间转译分析

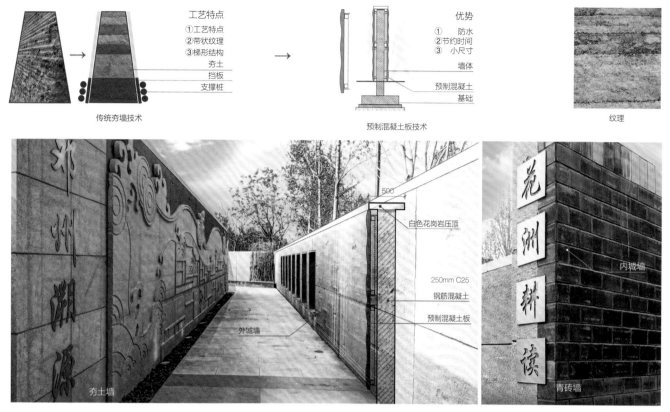

乡土材料的应用

植物应用展示

邓州八景图

项目展示

利用外城墙线性空间引导游客进入展园之中，景墙上的诗卷与地面雕刻的古地图共同构成了一幅立体长卷，展现着邓州的人文底蕴与自然风貌。

穰城寻迹

展示空间与交通空间

花洲耕读：以月季花岛的形式，再现花洲形态，通过地被月季、树状月季的搭配组合，以及流畅的铺装线条、细腻的场地铺装，展现"花洲"概念

古城今貌：展园内双城空间的层层跌水与逐级抬升的台阶形成了景观暗示，将园外的水景借于园内，青砖墙与夯土墙倒映在水面之上

诗意栖居的文化生活空间
——河北保定清泽园

Cultural Life Space of Poetic Dwelling
— Qingze Garden, Baoding City, Hebei Province

建成时间：2019 年	Time of completion: 2019
建设地点：河北省邢台市桥东区	Construction site: Qiaodong District, Xingtai City, Hebei Province
项目面积：0.85 hm²	Project area: 0.85 hm²
建设单位：保定市园林管理局	Construction unit: Baoding Landscape Administration Bureau
获奖信息：2020 年国际风景园林师联合会亚非中东地区文化与传统类荣誉奖	Awards: Award International Federation of Landscape Architects Asia-Africa Middle East Region Award (IFLA AAPME) Culture and Traditions Honorable Mention in 2020
项目团队：李雄 肖遥 郑小东 邵明 等	Project team: LI Xiong, XIAO Yao, ZHENG Xiaodong, SHAO Ming, et al

清泽园建于第三届中国河北省园林博览会，面积 8500m²。项目以保定自然山水和城市蓝绿肌理为概念主体，利用现代弹性低碳新材料"竹钢"对中国河北省 3000 年非物质文化遗产进行转译表述，以促使人们更多的关心和了解传统的非物质文化遗产。

与传统模式不同，本项目以新材料的参数化设计为载体表达传统美学与文化遗产，串联并阐释设计内容的古今内涵。项目的建设过程既是对中国非物质文化遗产的发掘又是对其保护和传承的积极实践，而新材料的使用也有效降低了项目成本，对保护和传承中国非物质文化遗产起到了巨大作用。

本项目最终以新型低碳弹性材料"竹钢"组成结构单元，通过参数化的方式将内部格栅、竹制墙体、竹筒填充等形式进行组合变化，通过六个区域的主题展示和文化提取，充分遵循乡土化、生活化的设计理念，形成以水景空间为特色，以文化遗产展示为内涵的独特城市展园景观，最终实现自然基底和人居理想空间的完美重合。

项目借助园林博览会的平台，推动了以弹性材料对非物质文化遗产转译表达的设计模式，产生了直接的社会效益和间接的文化价值。它不仅对文化遗产的保护发扬有着深远的影响，也为社会各界提供了文化保护和材料创新的新思路。

Qingze Garden was built in the 3rd China Hebei Garden Expo with an area of 8500 square meters. The project takes the Baoding natural landscape and city blue-green texture as the main concept, and uses the modern flexible low-carbon new material "bamboo steel" to translate and express the 3000-year intangible cultural heritage of Hebei Province, China, so as to encourage people to pay more attention to and understand the traditional intangible cultural heritage.

Different from the traditional model, this project uses the parametric design of new materials as the carrier to express the traditional aesthetics and cultural heritage, and to connect and explain the ancient and modern connotations of the design content. The construction process of the project is not only the exploration of China's intangible cultural heritage, but also the active practice of its protection and inheritance. The use of new materials also effectively reduces the project cost, which plays a huge role in the protection and inheritance of China's intangible cultural heritage.

This final project by new low-carbon elastic material of the bamboo steel structure unit, by means of parametric internal grille, bamboo wall, bamboo tube filling combination forms, such as through the theme of the six regional features and culture, fully follow the design concept of local-color, adaptation, waterscape space form to the characteristics, cultural heritage demonstrated as the connotation of unique urban exhibition park landscape, achieve the ideal of natural base and living space of perfect coincidence.

With the help of China Expo, the project promoted the design mode of heritage conservation driven by landscape architecture with elastic materials, and produced direct social benefits and indirect cultural values. It not only has a profound and long-term influence on the protection and development of cultural heritage, but also provides new ideas for cultural protection and material innovation for all sectors of the society.

本项目为中国河北省第三届园林博览会中的保定展园,位于中国河北省邢台市,面积约 8500m²。保定作为中国京津冀地区最古老的城市之一,具有超过 3000 年的历史,其间存留下不计其数的宝贵非物质文化遗产,而遗憾的是,很长一段时间内,城市将重心放在经济发展上,绝大多数的传统文化遗产都湮没在历史中,无法被人铭记。

1. 主入口
2. 琴起明台
3. 书画映廊
4. 金秋槐林
5. 棋落闲亭
6. 花之香径
7. 漫堤寻芳
8. 幽潭叠瀑
9. 诗酒轩居
10. 竹林佳酿
11. 醉黔小径

总平面

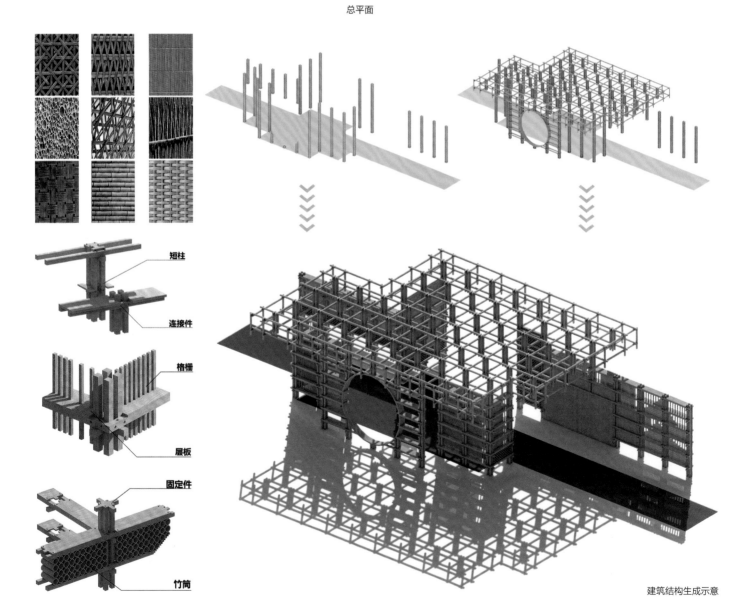

建筑结构生成示意

7　文化展园与艺术花园　247

主入口构筑

棋落闲亭

景墙　　　　　　　　　　　　　　　　　　　　　书画映廊

非物质文化遗产在现代风景园林中的传承与推广

中国河北省第三届园林博览会作为行业内具有代表性的交流活动和专业展览，在举办期间吸引了近 200 万人次的参观者。

保定园的设计内容直观地展示了当地的文化遗产和传统文化，取得了良好的景观效果，展示了具有地域特色的传统文化，对提升城市形象起到了重要作用。

同时，借助园博会平台，广泛推动了以新材料构筑为主导的遗产保护设计，并带来了直接的社会效益。保定园设计完成至今，保定市已经陆续开展了一系列文化公园的新建活动，并开展了一系列的展览和培训，让数以百万的市民对保定市 3000 年传承的非物质文化遗产有了新的认识和了解，这将对中国传统文化的未来传承与发展产生深远影响。

竹钢构筑

南湖菱秀：利用竹、苇、菊等低成本、低维护的乡土野趣植物，结合自然水系，打造富有自然野趣的中国传统文化意象

竹钢构筑：设计建造展示出现代材料创新应用的可能性

西溪芦影

后　记

当代生态哲学中，将人们对环境和自然的态度分为两种：一种是以人的利益为目标的自然观，另一种是以整个生态系统及其存在物（包括人类）的利益为目标的自然观。前者属于浅层的生态观，后者属于深层的生态观。显然，生态文明的理念是后一种观点。它多从生物圈的角度出发来建设，关注的是每个物种和生态系统的生存条件，而不是把注意力完全集中于它对人类价值作用方面。这正是生态文明所倡导的"尊重自然"的理念。

随着我国城市化进程的加快，城市人居环境中的问题越来越突出，园林绿地作为城市中唯一具有生命力的"绿色基础设施"，逐渐成为城市可持续发展的主力。园林绿地除了担负着缓解城市中的环境问题的功能外，还具有保障城市的生态安全、保护和维持生物多样性等重要功能，由此成为生态文明建设的重要场所。

在我国构建和谐社会的当下，尤其是党的十九届五中全会将"生态文明建设实现新进步"作为"十四五"时期经济社会发展主要目标之一的今天，风景园林的建设被赋予了新的内涵。一次又一次的风景园林实践探索让我们越发意识到风景园林的建设要旨应放在整个生态系统上，瞄准生物多样性的保护，才能真正实现人与自然的和谐共处，才能最终使人居环境走上健康的可持续发展之路，而这样的城市和风景园林也才具有生态文明的时代特征。

从教 30 年来，我主持了两百余项风景园林规划设计项目，获得了几十余项国际级、国家级、省部级规划设计奖励，从未停止探索生态文明背景下的风景园林实践。本书从这些项目中择取了 31 项风景园林实践作品，按照公园城市与国土生态空间、园林博览会与事件性景观、植物园、生态修复与郊野公园、海绵城市与滨水空间、城市更新与绿色开放空间、文化展园与艺术花园 7 种类型重新编排，对生态文明建设背景下不同类型风景园林实践的发展方向进行了总结与思考，全面展示了各个规划设计作品处理人与自然关系的策略与方法，以及发挥出的生态效益和社会价值。希望能引起风景园林设计师的共鸣，引导从业者树立深层的生态观念，从人类的长远利益出发，把整个生态系统作为关注对象，用整体的普遍联系的观点支撑设计理念，创作出"尊重自然、顺应自然、保护自然"的风景园林实践作品，为世界人居生态环境建设贡献中国方案和中国智慧。

在此，我要由衷地感谢陈俊愉院士、孟兆祯院士等老一辈先生们对我国风景园林事业作出的开拓性贡献。感谢我的研究生导师郦芷若、苏雪痕先生对我在风景园林学习与实践过程中的传道、授业、解惑。感谢行业内的各位同仁，谢谢你们的大力支持。

感谢郑曦、李运远、尹豪、蔡凌豪、冯潇、姚朋、李冠衡、张云路、戈晓宇、王鑫、李正、宋文、肖遥、林辰松、孙漪南、李方正、胡楠、马嘉、葛韵宇、邵明等各位老师及团队成员在参与实践项目过程中付出的辛苦。感谢本书收录的设计作品的建设单位。感谢为本书的编排做出贡献的老师和同学们。最后感谢中国建筑工业出版社杜洁、李玲洁编辑的鼎力支持。

习近平新时代生态文明思想为我国风景园林实践提供了系统的指引，同时也提出了新的要求，为延续人类文明产生了划时代的意义。风景园林作为实现生态文明的有效载体，在生态文明建设的过程中发挥重要的作用。风景园林从业者应不断创新理念和思维，主动响应国家战略和人居生态环境发展需求，不断深化城市园林的发展内涵，坚定不移地推动城市高质量发展，努力把我国风景园林建设有机统一于新时代中国特色社会主义建设的伟大实践中。

Postscript

In contemporary ecological philosophy, people's attitudes towards environment and nature are divided into two categories: one is the view of nature aiming at the interests of human beings, the other is the view of nature aiming at the interests of the whole ecosystem and its existence (including human beings). The former belongs to the shallow ecological view, while the latter belongs to the deep ecological view. Obviously, the idea of ecological civilization is the latter. It is constructed from the perspective of biosphere, focusing on the living conditions of each species and the ecosystem, rather than focusing on its role in human health. This is the concept of "respecting nature" advocated by ecological civilization.

With the acceleration of urbanization in China, the problems of urban living environment are becoming more and more prominent. As the only "green infrastructure" with vitality in the city, green space has gradually become the main force of urban sustainable development. In addition to the function of alleviating the environmental problems in the city, the green space also has the important functions of ensuring the ecological security of the city, protecting and maintaining the biodiversity, which has become an important place for the development of ecological civilization.

At the moment of building a harmonious society in China, especially when the Fifth Plenary Session of the 19th CPC Central Committee regards "new progress in the development of ecological civilization" as one of the main goals of economic and social development in the 14th Five Year Plan Period, the construction of urban landscape has been given a new connotation. The practice and exploration of landscape architecture again and again make us more aware that the construction of urban landscape architecture should focus on the whole ecosystem, aiming at the protection of biodiversity to truly achieve the harmonious coexistence of man and nature, and finally lead the development of human settlement environment in a healthy and sustainable manner. Such cities and urban green space would present the characteristics of ecological civilization.

In the past 30 years, I have presided more than 200 landscape planning and design projects, and won dozens of international, national, provincial, and ministerial planning and design awards. I have never stopped exploring landscape practice under the background of ecological civilization. This book selects 31 landscape architecture practice works from these projects, and rearranges them according to 7 types: park city and territorial ecological space, garden exposition and event landscape, botanical garden, ecological restoration and country parks, sponge city and waterfront space, urban renewal and green open space, and cultural garden and art garden, so as to summarize and consider the development direction of different types of landscape architecture practice under the background of ecological civilization development, and comprehensively show the strategies and methods of dealing with the relationship between man and nature, as well as the ecological benefits and social value of each planning and design works. I hope that it can arouse the resonance of landscape architects, guide practitioners to establish deep ecological concept, take the whole ecosystem as the object of concern for the long-term interests of mankind, support the design concept with the overall universal connection point of view, and create landscape works of "respecting nature, conforming to nature and protecting nature", so as to contribute China's proposals and China's wisdom to the construction of human settlements and ecological environment in the world.

Hereby, I would like to express my sincerest gratitude to academician Chen Junyu, academician Meng Zhaozhen and other senior landscape architects for their pioneering contributions to landscape architecture. I would like to thank my graduate tutors Li Zhiruo and Su Xuehen for their preaching, teaching and guidance in my study and practice of landscape architecture. My gratitude also goes to the industry peers for their great support.

I would thank Zheng Xi, Li Yunyuan, Yin Hao, Cai Linghao, Feng Xiao, Yao Peng, Li Guanheng, Zhang Yunlu, Ge Xiaoyu, Wang Xin, Song Wen, Xiao Yao, Lin Chensong, Sun Yinan, Li Fangzheng, Hu Nan, Ma Jia, Ge Yunyu, SHAO Ming and other teachers and team members for their hard work in participating in the project. I would also extend my thanks to the construction unit of the design works included in this book, to the teachers and students for their contribution to the arrangement of this book, and finally to Du Jie and Li Lingjie from China Architecture Publishing & Media Co., Ltd for their support.

Xi Jinping's thought of ecological civilization in the new era has provided systematic guidance for the practice of landscape architecture in China and has also put forward new requirements for epoch-making significance for the continuation of human civilization. As an effective medium of ecological civilization, urban landscape plays an important role in the process of ecological civilization. Landscape architecture practitioners should constantly innovate ideas and thinking, actively respond to the national strategy and the development needs of human settlement ecological environment, constantly deepen the development connotation of urban landscape architecture, unswervingly promote the high-quality development of cities, and strive to systematically integrate the construction of urban landscape architecture in the great practice of the construction of socialism with Chinese characteristics in the new era.

Li Xiong